Scalable Infrastructure for Distributed Sensor Networks

Krishnendu Chakrabarty and S. S. Iyengar

Scalable Infrastructure for Distributed Sensor Networks

With 109 Figures

 Springer

Krishnendu Chakrabarty, PhD
Department of Electrical and Computer Engineering, Duke University,
Box 90291, 120 Hudson Hall, Durham, NC 27708, USA

S.S. Iyengar, PhD
Department of Computer Science, Louisiana State University,
Baton Rouge, Louisiana 70803, USA

British Library Cataloguing in Publication Data
Chakrabarty, Krishnendu
    Scalable infrastructure for distributed sensor networks
    1. Sensor networks 2. Wireless LANs 3. Intelligent agents
    (Computer software)
    I. Title II. Iyengar, S. S. (Sundararaja S.)
    681.2
ISBN-10: 1852339519

Library of Congress Control Number: 2005924308

ISBN-10: 1-85233-951-9
ISBN-13: 978-1-85233-951-7
Springer Science+Business Media
springeronline.com

Typesetting: Electronic camera-ready by authors
Cover design: deblik, Berlin, Germany
Production: LE-TEX Jelonek, Schmidt & Vöckler GbR, Leipzig, Germany
Printed in Germany
69/3141-543210 Printed on acid-free paper SPIN 11329954

Prof. Iyengar dedicates this book to Dr. Narayana Murthy of Infosys for his contribution to the global development of IT industries, and to Prof.C.N.Rao for his contributions to many areas of basic science.

Preface

Advances in the miniaturization of microelectromechanical systems have led to battery-powered sensor nodes that have sensing, communication and processing capabilities. These sensor nodes can be networked in an ad hoc manner to perform distributed sensing and information processing. Such ad hoc sensor networks provide greater fault tolerance and sensing accuracy and are typically less expensive compared to the alternative of using only a few large isolated sensors. These networks can also be deployed in inhospitable terrains or in hostile environments to provide continuous monitoring and processing capabilities.

A typical sensor network application is inventory tracking in factory warehouses. A single sensor node can be attached to each item in the warehouse. These sensor nodes can then be used for tracking the location of the items as they are moved within the warehouse. They can also provide information on the location of nearby items as well as the history of movement of various items. Once deployed, the sensor network needs very little human intervention and can function autonomously. Another typical application of sensor networks lies in military situations. Sensor nodes can be air-dropped behind enemy lines or in inhospitable terrain. These nodes can self-organize themselves and provide unattended monitoring of the deployed area by gathering information about enemy defenses and equipment, movement of troops, and areas of troop concentration. They can then relay this information back to a friendly base station for further processing and decision making.

Sensor nodes are typically characterized by small form-factor, limited battery power, and a small amount of memory. Due to their limited resources, many of the methods developed for the Internet and mobile ad hoc networks cannot be directly applied to sensor networks. A scalable infrastructure is to solve problems such as data routing, self-organization and data dissemination that emerge in sensor networks. This book is focused on a scalable infrastructure for information processing in wireless sensor networks. It addresses the problems of coverage-centric sensor deployment, energy-efficient self-organization, target localization, information dissemination, data routing, and time synchronization.

Book Overview

Chapter 1 presents an overview of various networking and information processing issues related to sensor networks. Chapters 2–5 are based on recent research carried out at Duke University. Chapter 2 addresses cluster-based sensor networks. It presents the virtual force algorithm (VFA) as a new approach for sensor deployment to improve the sensor field coverage after an initial random placement of sensor nodes. The cluster head executes the VFA algorithm to find new locations for sensors to enhance the overall coverage. This chapter also considers the sensor deployment problem when unavoidable uncertainty exists in precomputed sensor node locations, e.g. for airdropped sensor nodes. Inherent uncertainties in the designated sensor positions as well as the sensor field terrain are integrated in the sensor deployment.

Chapter 3 describes an energy-aware target localization strategy based on a two-step communication protocol between the cluster head and the sensors reporting the target detection events. This approach reduces energy consumption in the target localization process for wireless sensor networks by making use of the existing information redundancy in the target data from sensor nodes. It also offers the built-in advantages of reducing the communication bandwidth and filtering out false alarms.

Chapter 5 describes a new energy-efficient flooding algorithm termed LAF for data dissemination in wireless sensor networks. The proposed d approach uses the concept of virtual grids to divide the monitored area and nodes then self-assemble into groups of gateway nodes and internal nodes. It exploits the location information available to sensor nodes to prolong the lifetime of a sensor network by reducing the redundant transmissions that are inherent in flooding.

Chapters 6–7 are based on work carried out at Louisiana State University. Chapter 6 presents three energy equivalence routing approaches to balance the network-wide energy consumption and prolong the lifetime of network. Finally, Chapter 7 presents a time synchronization method for dynamic and ad hoc sensor networks. Energy efficiency and scalability are important attributes of this method.

In summary, this book provides an important link between the crucial problems of coverage, connectivity, energy management, and self-organization in wireless sensor networks. It is expected to lead to even more efficient protocols for node deployment, state management, and information processing for energy-constrained and failure-prone sensor networks.

Acknowledgments

We are grateful to Yi Zou and Harshavardhan Sabbineni of Duke University for their contributions to Chapters 1–5 of this book. These chapters are based on the Ph.D. thesis of Yi Zou and the M.S. thesis of Harshavardhan Sabbineni. We also thank Wei Ding of Louisiana State University for contributing to Chapters 6–7. This book grew out of a research project supported by the Office of Naval Research and the Defense Advanced Research Project Projects Agency. Financial support received from these agencies is gratefully acknowledged. We thank Anthony Doyle, Engineering Editor at the London office of Springer-Verlag, for supporting this book project. We also thank Kate Brown at the London office Springer-Verlag for her timely help in the preparation and typesetting of this manuscript. Finally, we thank Andrea Koehler for help with last-minute corrections.

Krishnendu Chakrabarty
Durham, NC
S. S. Iyengar
Baton Rouge LA
June 2005

Contents

XIV   Contents

1

Introduction

Advances in miniaturization of microelectronic and mechanical structures (MEMS) have led to battery-powered sensor nodes that have sensing, communication and processing capabilities [39, 96]. These sensor nodes can be networked in an ad hoc manner to perform distributed sensing and information processing in many situations. Such ad hoc sensor networks provide greater fault tolerance and sensing accuracy and are typically less expensive compared to the alternative of using only a few large isolated sensors. These networks can also be deployed in inhospitable terrains or in hostile environments to provide continuous monitoring and processing capabilities for a wide variety of applications.

Wireless sensor networks that are capable of observing the environment, processing data, and making decisions based on these observations, have therefore attracted considerable attention recently [3, 11, 38, 39, 96, 98, 110, 120]. These networks are important for a number of applications such as coordinated target detection and localization, surveillance, and environmental monitoring. Breakthroughs in miniaturization, hardware design techniques, and system software have led to cheaper sensors and fueled recent advances in wireless sensor networks [2, 3, 11, 39, 110]. In a 1999 article titled *21 Ideas for the 21st Century* published in *Business Week*, Nobel Laureate Horst Stormer wrote, "Untethered micro sensors will go anywhere and measure anything—traffic flow, water level, number of people walking by, temperature. This is developing into something like a nervous system for the earth, a skin for the earth. The world will evolve this way".

A sensor network consists of a large number of nodes, referred to as sensor nodes. A sensor node integrates hardware and software for sensing, data processing, and communication. Sensor nodes can be deployed readily in large number for various types of environments. They rely on wireless channels for transmitting data to and receiving data from other nodes. The maximum distance that a node can communicate with another node is characterized by the communication unit on the sensor node, e.g., for the RF sensors used in the Berkeley mote [10], the maximum operation range is approximately up to 100 ft. The sensing area of a sensor node depends on the type of physical sensors used on that node, e.g., a range sensor such as the Polaroid 6500 ultrasonic ranging module, commonly used in robotics applications, is able to detect a target from 6 inches away up to a distance of 35 feet [94]. Figure 1.1 illustrates the basic structure of a sensor node. The attributes of some commercially available nodes are listed in Table 1.1. An important consideration in sensor networks is the amount of energy required for sensing, computation, and communication. The lifetime of a sensor node depends to a large extent on the battery lifetime, hence it is extremely important to adopt energy-efficient strategies for information processing.

Table 1.1. Specifications of inexpensive wireless sensors, as provided on the website www.xbow.com/Products/Wireless_Sensor_Networks.htm.

Model	MICA2DOT	MICA2	MICAz
Battery	3V coin cell	AA × 2	AA × 2
Size (mm)	25 × 6	58 × 32 × 7	58 × 32 × 7
Weight (g)	3	18	18
Range (m)	150	150–300	75–100
Data rate	38.4 KBaud	38.4 KBaud	250 Kbps
Program flash memory	128 KB	128 KB	128 KB
Serial flash memory	512 KB	512 KB	512 KB
EEPROM	4 KB	4 KB	4 KB
RF (MHz)	315/433/868/916	315/433/868/916	2400

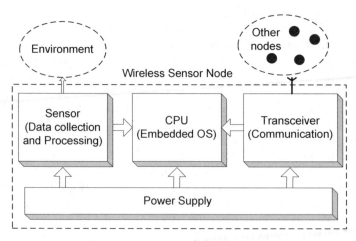

Fig. 1.1. Sensor node architecture.

A typical sensor network application is inventory tracking in factory ware-houses. As illustrated in Fig. 1.2, a single sensor node can be attached to each item in the warehouse. These sensor nodes can then be used for tracking the location of the items as they are moved within the warehouse. They can also provide information on the location of nearby items as well as the history of movement of various items. Once deployed, the sensor network needs very little human intervention and can function autonomously. Another typical application of sensor networks lies in military situations. Sensor nodes can be air-dropped behind enemy lines or in inhospitable terrain. These nodes can self-organize themselves and provide unattended monitoring of the deployed area by gathering information about enemy defenses and equipment, move-ment of troops, and areas of troop concentration. They can then relay this information back to a friendly base station for further processing and decision making. This is illustrated in Fig. 1.3, where the presence of an enemy tank in the monitored area is relayed to the command center.

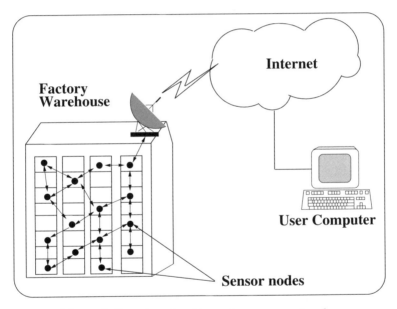

Fig. 1.2. Inventory tracking using sensor networks.

In recent years, wireless sensor networks have also been deployed for a number of applications. In the spring of 2002, the Intel Research Laboratory at Berkeley initiated a collaboration with the College of the Atlantic in Bar Harbor and the University of California at Berkeley to deploy wireless sensor networks on Great Duck Island, Maine [80, 118]. These networks monitor the microclimates in and around nesting burrows used by the Leach's Storm Pe-

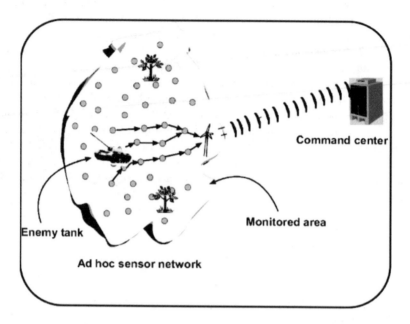

Fig. 1.3. Battlefield monitoring using sensor networks.

trel. The goal of this project is to develop a habitat monitoring kit that enables researchers worldwide to engage in the non-intrusive and non-disruptive monitoring of sensitive wildlife and habitats (www.greatduckisland.net/index.php).

At the end of the field season in November 2002, well over 1 million readings had been logged from 32 motes deployed on the island. Each mote was equipped with a microcontroller, a low-power radio, memory and batteries. For habitat monitoring, sensors were added for temperature, humidity, and barometric pressure. Motes periodically sampled and relayed their sensor readings to computer base stations on the island. These in turn fed into a satellite link that allowed researchers to access real-time environmental data over the Internet.

In June 2003, a second generation network with 56 nodes was deployed. The network was augmented in July 2003 with 49 additional nodes and again in August 2003 with over 60 more burrow nodes and 25 new weather station nodes. These nodes form a multihop network transferring their data back "bucket brigade" style through dense forest. Some nodes are more than 1000 feet deep in the forest providing data through a low power wireless transceiver.

In another project, on Aug. 27, 2001, researchers from the University of California, Berkeley and the Intel Berkeley Research Lab demonstrated a self-organizing wireless sensor network consisting of over 800 tiny low-power sensor nodes (today.cs.berkeley.edu/800demo). It utilized tiny nodes, just the size of a quarter, that were hidden under 800 chairs in the lower section of a presentation hall. The core of a node was a 4 MHz low power microcontroller (ATMEGA 163) providing 16 KB of flash instruction memmory, 512 bytes of SRAM, ADCs, and primitive peripheral interfaces. A 256 KB EEPROM served as secondary storage. Sensors, actuators, and a radio network served as the I/O subsystem. The network utilized a low-power radio (RF Monolithics T1000) operating at 10 kbps.

As illustrated in Figure 1.1, there are several major issues underlying the design and implementation of wireless sensor networks, including sensing, networking, data dissemination and self-organization , embedded computing, and power management. Since existing techniques and methodologies for these four topical areas cannot be directly applied to wireless sensor networks, research in this area is difficult and challenging. The sensor data collected in a distributed manner must be aggregated to obtain a coherent and global understanding of the environment [18, 59, 63]. Information exchange and data dissemination must be carried out using efficient communication protocols. A number of system architectures, communication protocols, and data aggregation algorithms have been proposed in the literature for retrieving and processing sensed data with low energy consumption [40, 51, 53, 54, 136, 137]. Wireless sensor networks are typically organized in an ad hoc manner, e.g., through random sensor deployment and ad hoc networking protocols. Nevertheless, a number of methods have recently been developed to organize the sensor network in hierarchical clusters [10, 86, 117] to improve the sensing coverage [23, 29, 134, 135] and reduce the energy consumption in information processing [52, 76, 77, 136, 137].

1.1 Challenges

In this section, we motivate the challenges involved in the design of self-organization and data dissemination protocols in wireless sensor networks. Due to their resource constraints and unique application requirements, sensor networks pose a number of challenges. These are summarized below:

- **Small Memory:** Sensor nodes usually have a small amount of memory. Hence, sensor network protocols should not require the storage of a large amount of information at the sensor node.
- **Limited Battery Power:** Sensor nodes typically have a small form factor with a limited amount of battery power [86]. Furthermore, radio communication typically costs more in terms of energy compared to computation costs in a sensor node. Therefore, protocols designed for sensor networks should utilize only a few control messages.

- **Fault Tolerance:** Sensor nodes are prone to failure. This may be due to a variety of reasons. Loss of battery power may lead to failure of the sensor nodes. Similarly, when sensor nodes are deployed in hostile or harsh environments as in the case of military or industrial applications, sensor nodes might be easily damaged. Thus, protocols designers should build fault tolerance into their algorithms for improving the utility of sensor networks.
- **Self-Organization:** Sensor nodes are often air-dropped in hostile or harmful environments. It is not possible for humans to reach these sensor nodes. Besides, it is not possible for humans to repair each sensor node, as often the number of sensor nodes is quite large. Hence, self-organization of sensor nodes to form a connected network is an essential requirement.
- **Scalability:** The number of sensor nodes in a sensor network can be in the order of hundreds or even thousands. Hence, protocols designed for sensor networks should be highly scalable.

1.2 Sensor Network Architectures

This section discusses different sensor network architectures. A typical sensor node consists of four basic components: a power unit that may be battery-powered, a sensing unit that may consist of one or more sensors, a processing unit that consists of a CPU to provide a basic processing capabilities, a DSP chip to provide limited signal processing functions, and a transceiver to provide untethered communications.

Sensor network applications such as inventory tracking, perimeter defense, and environmental monitoring require careful planning in the design of the protocols, including the choice of the sensor network architecture and the amount of redundancy to be present in the network. There are several sensor network architectures that protocol designers might consider for their applications:

- *Homogeneous versus heterogeneous:* A sensor network may consist of homogeneous or heterogeneous nodes. In a homogeneous sensor network, all the sensor nodes have similar sensing and processing abilities. As a typical sensor network can have up to thousands of nodes, homogeneous sensor networks are economical due to reasons of scale. A heterogeous sensor network may consist of sensor nodes with different sensor types, power capacities and processing abilities. An example of a heterogeneous sensor network is a habitat monitoring network where sensor nodes with cameras perform the video sensing while sensor nodes with recorders perform audio sensing, both with different power requirements and processing abilities. Thus, different protocols are needed for homogeneous and heterogeneous sensor networks.
- *Random versus deterministic deployment:* Sensor nodes can be deployed by air-dropping them (Fig. 1.4) or throwing them randomly in a target area

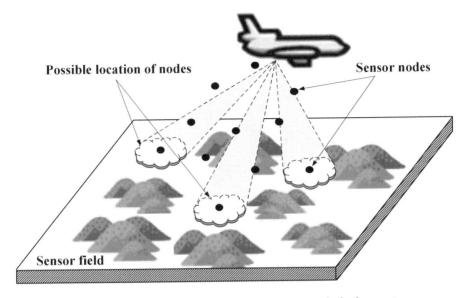

Fig. 1.4. An example of ad hoc sensor network deployment.

or they can be placed at pre-determined locations using a deterministic scheme. Protocols for self-configuration of a randomly-deployed network may not be well suited for a deterministically-deployed sensor network. Similarly, data dissemination algorithms designed for deterministic sensor networks may not perform well when used in randomly-deployed sensor networks.

- *Hierarchical versus flat topology:* Designers can select either a flat topology or a hierarchical cluster-based topology depending on the application for their protocol. Hierarchical topologies are generally more suited for sensor networks as they allow data fusion and other common functions within a cluster, thus minimizing communication outside a cluster. A three-level hierarchical sensor network is shown in Fig. 1.5.
- *Static versus mobile:* Sensor networks can consist of either static or mobile nodes, or a mixture of both static and mobile nodes. Depending on the composition of the particular sensor network, they may require very different algorithms for self-organization and data dissemination protocols.

1.3 Sensor Node Deployment

Sensor node deployment problems have been studied in a variety of contexts [18, 24, 59, 98, 122]. In the area of adaptive beacon placement and spatial localization, a number of techniques have been proposed for both fine-grained

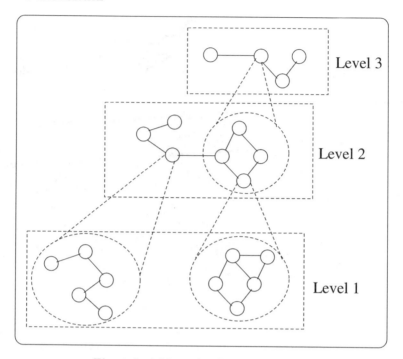

Fig. 1.5. A hierarchical sensor network.

and coarse-grained localization [20, 50]. Sensor deployment and sensor planning for military applications are described in [88], where a general sensor model is used to detect elusive targets in the battlefield. The sensor coverage analysis is based on a hypothesis of possible target movements and sensor attributes. However, the proposed wireless sensor networks framework in [88] requires a considerable amount of *a priori* knowledge about possible targets. A variant of sensor deployment has been considered for multi-robot exploration [28, 57, 79]. Each robot can be viewed as a sensor node in such systems. An incremental deployment algorithm is used in which sensor nodes are deployed one by one in an adaptive fashion. Each new deployment of a sensor is based on the sensed information from sensors deployed earlier. A drawback of this approach is that it is computationally expensive. As the number of sensors increases, each new deployment results in a relatively large amount of computation.

In [54], the concept of potential force is used in a distributed fashion to perform sensor node deployment in ad hoc wireless sensor networks. The problem of evaluating the coverage provided by a given placement of sensors is discussed in [83, 84]. The major concern here is the self-localization of sensor nodes; sensor nodes are considered to be highly mobile and they move frequently. An optimal polynomial-time algorithm that uses graph theory and computational geometry constructs is used to determine the best-case and the worst-case coverage. Radar and sonar coverage also present several related challenges [97]. Radar and sonar netting optimization are of great importance for detection and tracking in a surveillance area. Based on the measured radar cross-sections and the coverage diagrams for the different radars, the authors in [97] propose a method for optimally locating the radars to achieve satisfactory surveillance with limited radar resources.

Sensor placement on two- and three-dimensional grids has been formulated as a combinatorial optimization problem, and solved using integer linear programming in [23, 24]. This approach suffers from two main drawbacks. First, computational complexity makes the approach infeasible for large problem instances. Second, the grid coverage approach relies on "perfect" sensor detection, i.e. a sensor is expected to yield a binary yes/no detection outcome in every case. However, because of the inherent uncertainty associated with sensor readings, sensor detection must be modeled probabilistically [33, 34]. It is well known that there is inherent uncertainty associated with sensor readings; hence sensor detections must be modeled probabilistically. A probabilistic optimization framework for minimizing the number of sensors for a two-dimensional grid has been proposed recently [33, 34]. This algorithm attempts to maximize the average coverage of the grid points.

There also exists a close resemblance between the sensor placement problem and the art gallery problem (AGP) addressed by the art gallery theorem [106]. The AGP problem can be informally stated as that of determining the minimum number of guards required to cover the interior of an art gallery. (The interior of the art gallery is represented by a polygon.) The AGP has been solved optimally in two-dimensional and shown to be NP-hard in the 3-dimensional case. Several variants of AGP have been studied in the literature, including mobile guards, exterior visibility, and polygons with holes.

A related problem in wireless sensor networks is that of spatial localization [50, 97]. In wireless sensor networks, nodes need to be able to locate themselves in various environments and on different distance scales. Localization is particularly important when sensors are not deployed deterministically e.g., when sensors are thrown from airplanes in a battlefield, and for underwater sensors that might move due to drift. Sensor networks also make use of spatial information for self-organization and configuration. A number of techniques for both fine and coarse-grained localization have been proposed [20, 21].

Other related work includes the placement of a given number of sensors to reduce communication cost [64], and optimal sensor placement for a given target distribution [93]. Sensor deployment for collaborative target detection is discussed in [29], where path exposure is used as a measure of the effectiveness of the sensor deployment. This method uses sequential deployment of sensors, i.e., a limited number of sensors are deployed in each step until the desired minimum exposure or probability of detection of a target is achieved. In most practical applications however, we need to deploy the sensors in advance without any prior knowledge of the target and sequential deployment is often infeasible. Moreover, sequential deployment may be undesirable when the number of sensors or the area of the sensor field is large. Thus a single step-deployment scheme is more advantageous in such scenarios. In [77], the authors propose a dual-space approach to event tracking and sensor resource management.

In this book, we describe a virtual force algorithm (VFA) as a sensor deployment strategy to enhance the coverage after an initial random placement of sensors. The VFA algorithm is based on disk packing theory [78] and the virtual force field concept from physics and robotics [28, 57, 79]. For a given number of sensors, the VFA algorithm attempts to maximize the sensor field coverage. A judicious combination of attractive and repulsive forces is used to determine the new sensor locations that improve the coverage. Once the effective sensor positions are identified, a one-time movement with energy consideration incorporated is carried out, i.e. the sensors are redeployed, to these positions. The sensor field is represented by a two-dimensional grid. The dimensions of the grid provide a measure of the sensor field. The granularity of the grid, i.e., distance between grid points can be adjusted to trade off computation time of the VFA algorithm with the effectiveness of the coverage measure. The detection by each sensor is modeled as a circle on the two-dimensional grid. The center of the circle denotes the sensor while the radius denotes the detection range of the sensor. We first consider a binary detection model in which a target is detected (not detected) with complete certainty by the sensor if a target is inside (outside) its circle. The binary model facilitates the understanding of the VFA model. We then investigate realistic probabilistic models in which the probability that the sensor detects a target depends on the relative position of the target within the circle.

We also formulate an uncertainty-aware sensor deployment problem to model scenarios where sensor locations are precomputed but the sensors are airdropped or dispersed. In such scenarios, sensor nodes cannot be expected to fall exactly at predetermined locations; rather there are regions where there is a high probability of sensor being actually located. Such examples include airdropped sensor nodes and underwater sensor nodes that drift due to water currents. Thus a key challenge in sensor deployment is to determine an uncertainty-aware sensor field architecture that reduces cost and provides high coverage, even though the exact location of the sensors may not be completely controllable. In this book, we present two algorithms for sensor deployment wherein we assumed that sensor positions are not exactly predetermined. We assume that the sensor locations are calculated before deployment and an attempt is made during the airdrop to place sensors at these locations; however, the sensor placement calculations and coverage optimization are based on a Gaussian model, which assumes that if a sensor is intended for a specific point P in the sensor field, its exact location can be anywhere in a "cloud" surrounding P. Note that the placement algorithms give us the sensor positions prior to actual placement and we assume that sensors are deployed in a single step.

1.4 Energy-Efficient Information Processing

Energy management in sensor networks is crucial since battery-driven sensor nodes are severely energy-constrained. Considerable research has been recently carried out in an effort to make sensor networks energy-efficient [12, 52, 76, 81, 86, 111, 112, 123, 136, 137]. In [12, 13], a mathematical model is presented to determine a bound on sensor network lifetime, with and without sensing activities. The hardware-based energy model for transmission and reception for the wireless sensor node described in [104] is widely used as the basic communication energy consumption model for a wireless sensor network node. In [51], a cluster-based routing algorithm called LEACH is proposed as part of an energy-efficient communication protocol for wireless sensor networks. The self-selected cluster heads collect raw data from the neighboring sensing nodes, aggregate them by data fusion methods, and transmit the aggregated data back to base stations for higher-level processing. PEGASIS, an improvement over LEACH, is another example of an energy-aware protocol [76], which tends to increase the sensor network lifetime by decreasing the bandwidth via local collaboration among nodes. Another example of an energy-efficient approach is the TEEN protocol proposed in [81]. Dynamic power management [112] has also been used for the design of energy-efficient wireless sensor networks. Other related work includes energy-saving strategies for the link layer [111], data aggregation [123], and system partitioning [124]. In [113], the authors propose a heuristic power-efficient sensor deployment strategy by selecting mutually-exclusive sensor sets to cover the sensor field. The sensors in only one set are awake at a time; the others are asleep. An energy-efficient strategy for target localization is proposed in [136, 137].

Our focus here is on reducing energy consumption in wireless sensor networks for target localization and data communication. In general, a sensor network has an almost constant rate of energy consumption if no target activities are detected in the sensor field [12, 13]. The minimization of energy consumption for an active sensor network with target activities is more complicated since target detection involves collaborative sensing and communication involving different nodes. The transmission of detailed target information consumes a significant amount of energy due to the large volume of raw data. Contention for the limited bandwidth among the shared wireless communication channels causes additional delay in the relaying of detailed target information to the cluster head. In this book, we show how we can prolong the sensor network lifetime by adopting a new target localization procedure. We propose an *a posteriori* energy-aware target localization strategy, which is based on a two-step communication protocol between the cluster head and the sensors reporting the target detection events. In the first step, sensors detecting a target report the event to the cluster head using a very short binary yes/no message. The cluster head subsequently queries a subset of sensors that are in the vicinity of these likely target positions. This subset is determined from the localization procedure executed by the cluster head. Preliminary simulation results show that a large amount of energy is saved by using the proposed target localization procedure. These results also illustrate the built-in advantages of the proposed target localization procedure in reducing the communication bandwidth and filtering out false alarms.

1.5 Data Dissemination Algorithms

Once sensor nodes are deployed, efficient protocols are needed to disseminate the data sensed by the sensor nodes. Data dissemination involves the sending of the sensed data to the nodes that requested the data from the area where the event has occurred. As many sensors might be present in the area of event occurrence, data might be duplicated and nodes may receive multiple copies, or data may be lost due to a lossy communication channel. Several schemes have been proposed for data dissemination in sensor networks. In this section, we describe the state-of-the-art data dissemination protocols for wireless sensor networks.

Traditionally, flooding is used in networks to disseminate information. It is also used in several routing algorithms in sensor networks [129]. In flooding, the source node broadcasts the packet to all its neighbors. Each node that receives the packet stores a copy of the packet and broadcasts the packet to all its neighbors. Flooding terminates when a maximum number of hops are reached or the destination of the packet is the node itself.

Flooding is robust to node failures and delivers the packet to all the nodes in the network provided the network is lossless. However, the following problems might exist if flooding is done indiscriminately. As each node may be in the transmission range of many other nodes, each node might receive multiple copies of the same packet, thereby resulting in wastage of energy. Similarly, as sensor networks are typically very dense, heavy contention might result because many nodes are trying to acquire the channel at the same time. These problems are collectively referred to as the broadcast storm problem [14]. In addition to receiving redundant packets mentioned above, another problem might occur in wireless sensor networks. If the packet received by a node already has some or all of the data contained in the packet, wastage of energy occurs. This is known as the overlap problem.

Gossiping [91] is another data dissemination protocol traditionally used in ad hoc networks. In gossiping, the source node sends the packet to a randomly-selected neighbor. Each node that receives the packet randomly selects a neighbor and sends the packet to it. This process is repeated by all the nodes that have received the packet. Thus, data is disseminated throughout the network. In flooding, when a node with high degree receives a packet, it broadcasts the packet to all its neighbors In turn, all the neighbors broadcast new copies of the packet. Thus, the network is flooded with multiple copies of the same packet. Gossiping avoids the implosion problem by sending the packet to only one neighbor. This also results in energy savings. However, the information is also disseminated at a slower rate compared to flooding. Note that gossiping does not solve the overlap problem in sensor networks.

SPIN [66] protocols are a set of resource-adaptive information dissemination protocols for wireless sensor networks. In SPIN, nodes use meta-data to describe the data they possess. Nodes negotiate through a set of protocols to request the data they do not possess. When a node obtains new data, it broadcasts a ADV message to all of its neighbors with the meta-data describing the new data. Nodes that have received the ADV message checks the meta-data to see if it already has the data. Otherwise, it sends a REQ message to the sender of the ADV message requesting the data. The sender responds with a DATA message containing the requested data and the protocol terminates.

SPIN achieves energy savings by eliminating requests for redundant transmissions of data. Upon receipt of a ADV message, a node need not send a REQ message if it already has the data. Similarly, a node can aggregate its data with the newly received data and send an ADV message for the aggregated data. Nodes are also resource-adaptive in SPIN. Nodes poll their system resources for the amount of remaining energy and make informed decisions about disseminating information. This is shown in Fig. 1.6, where node A advertises its data by broadcasting an ADV packet. Nodes B and C respond by requesting the data using a REQ packet.

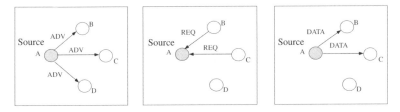

Fig. 1.6. Information dissemination in the SPIN protocol

GEAR [131] is a recursive data dissemination protocol for wireless sensor networks. A target region is specified in each query packet. Initially, the query packet is forwarded towards the target region. GEAR uses a set of geographically informed heuristics to route packets to the target region. Once the packet reaches the target region, it then uses a recursive geographic forwarding scheme to disseminate the packet within the target region. Each node within the target region splits its region into subregions and sends a copy of the query packet to each of the subregions. This recursive splitting and forwarding is terminated when a node is the only one in the subregion. The splitting of a region into subregions in recursive geographic forwarding is illustrated in Fig. 1.7.

Information dissemination in sensor networks is made information-aware in [32]. Every data packet is assigned a priority level based on its information content and criticalness. Packets carry a small amount of state to help make

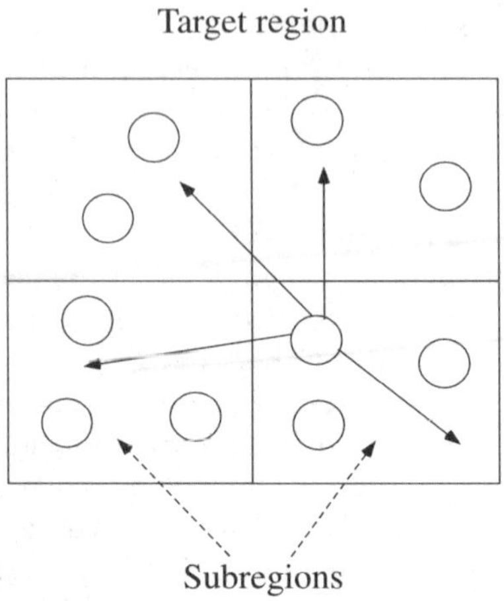

Target region
Subregions

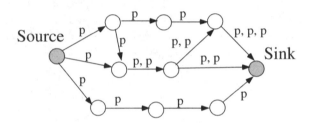

Fig. 1.7. Information dissemination in the recursive geographic protocol.

Source
Sink
p
p, p
p, p, p

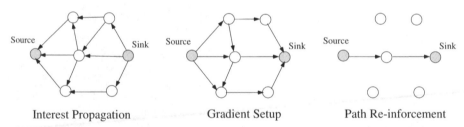

Fig. 1.8. Information dissemination in the multi-path forwarding protocol.

Source
Sink
Interest Propagation
Gradient Setup
Path Re-inforcement

Fig. 1.9. Information dissemination in the directed diffusion protocol.

forwarding decisions at individual nodes. The ReInForM technique in [32] uses local knowledge of channel error rates and neighborhood at each node. It provides the desired amount of reliability in data delivery by sending multiple copies of the same packet through multiple paths from the source to the sink. Multi-path forwarding from a source to a sink is illustrated in Fig. 1.8. A total of four paths exist between the source and the sink. The number of packets sent to the sink through each path assuming the communication links are lossless is shown in the figure.

Directed diffusion [129] is a data-centric paradigm for disseminating information. In Directed diffusion, data is named using attribute-value pairs. Query for a sensing task is distributed throughout the network as an interest for named data. This dissemination sets up gradients within the network to draw events matching the interest. The sensor network re-inforces a small number of these paths. Fig. 1.9 shows the three steps involved in directed diffusion, including interest propagation from the sink to the source, the setup of gradients from source to the sink, and the re-inforcement of better paths from the source to the sink. Directed diffusion positively re-inforces certain paths and negatively others to repair the paths that have failed nodes in them. It is a reactive routing technique and enables in-network aggregation of data by application-specific filters at each node in the network.

A qualitative comparison of different data dissemination protocols is shown in Table. 1.2.

Table 1.2. Comparison of different data dissemination protocols.

Data-Dissemination Protocol	No Overlap?	No Implosion?	Resource-Adaptive?	Fault Tolerance
Flooding	No	No	No	High
Gossiping	No	No	No	Low
SPIN	Yes	Yes	Yes	High
GEAR	Can be incorporated	Can be incorporated	Yes	Medium
Directed Diffusion	Yes	No	Yes	High
ReInForM	No	No	Can be incorporated	High

1.6 Self-Configuration Methods

Several sensor network applications require unattended autonomous operation for extended periods of time. Hence, sensor nodes should self-organize themselves and perform data gathering and processing in spite of node failures, loss of temporary communication links and node movement. In this section, a number of self-organization protocols in ad hoc sensor networks are discussed.

Span [26] attempts to save energy by switching off redundant nodes without affecting the connectivity of the network. In Span, a limited set of nodes self-organize themselves to form a multi-hop forwarding backbone while other nodes go to sleep. Nodes make decisions based on their local topology information.

A TDMA-based self organization scheme for sensor networks is presented in [114]. Each node uses a superframe, similar to a TDMA frame, to schedule different time slots for different neighbors. In each slot, a node can only talk

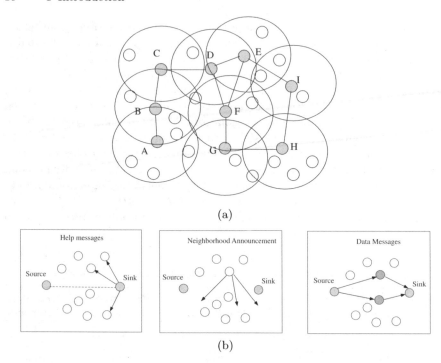

(a)

(b)

Fig. 1.10. Self-organization in sensor networks: (a) Connected sensor cover; (b) ASCENT.

to that neighbor for which the slot is reserved. Either code division multiple access (CDMA) or frequency division multiple access (FDMA) is used to prevent collision of packets among potentially interfering links. However, this scheme does not take advantage of the redundancy inherent in wireless sensor networks to power off some nodes.

The work in [47] introduced the concept of connected sensor cover for self-organizing the sensor network to achieve energy savings. A connected sensor cover for a query is the minimum set of sensors such that the sensing region of the selected sensors covers the entire geographical region of the query and the selected set of sensors form a connected communication graph. At each stage, the algorithm selects a path of sensors that connects an already connected sensor to a partially connected sensor. The selected path is then added to the already selected sensors at that stage. The algorithm terminates when the selected set of sensors completely cover the query region. This is illustrated in Fig. 1.10(a), where nodes A, B, C, D, E, F, G, H, and I completely cover the sensor field and form a connected sensor cover. This algorithm results in self-organization of the sensor nodes for a specific query to improve the energy efficiency of the sensor network.

A self-organizing algorithm termed *Rapid*, for message-efficient clustering based on the concept of budget allocation is presented in [65]. *Rapid* produces uses very few messages to produce clusters of bounded size. A node that wants to build the cluster initiates the process by allocating a certain budget to itself and and broadcasts the remaining budget among its neighbors by sending each neighbor a message. Nodes that receive the budget account for themselves and distribute the remaining budget among themselves. Each node that receives a message sends an acknowledgment to its parent when either the budget is exhausted or it has received acknowledgments from all its children. The algorithm terminates when the node that initiated the algorithm receives acknowledgments from all the neighbors it sent a message to.

The *Rapid* algorithm always produces clusters less than the desired limit. However, sometimes the produced cluster may be quite small compared to the desired bound. This is improved in *Persistent*, where a node does not immediately send an acknowledgment to its parent on receiving acknowledgments from all its children. It checks to see if the budget allocated to each of its neighbors has been exhausted. If the budget is not exhausted, it distributes the remaining budget among the neighbors it has not previously explored. Thus a node returns an acknowledgment only when its allocated budget has been met or if further growth is not possible. These algorithms produce single clusters of bounded size using few messages.

ASCENT [22] is a self-organizing scheme that provides topology control for sensor networks. In ASCENT, each node assesses its connectivity and adapts its participation in the multi-hop network. ASCENT has several phases. Upon initialization, each node enters into a listening-only phase called the *neighbor discovery phase*, where each node obtains an estimate of the number of neighbors actively transmitting based on local measurements. At the end of this phase, nodes enter into the join decision phase, where they decide whether to join the multi-hop sensor network. During this phase, a node may temporarily join the network for a certain period of time to check if contributes to improved connectivity. If a node decides to join the network for a longer the network for a longer time, it enters into an active phase and participates in the routing protocols of the network. If a node decides not to join the network, it enters into an adaptive phase where it turns itself off for a period of time or reduces its transmission range. Fig. 1.10(b) illustrates the various phases involved in ASCENT. The sink node A sends help messages in the *neighbor discovery phase* that results in neighborhood announcement by other nodes in the *join decision phase*. Nodes that have joined the network are in the active phase and participate in forwarding data from the source to the sink. ASCENT improves the energy-efficiency of the sensor network without a significant improvement in message loss. It is also adaptive to the traffic in the network and is stable under varied network traffic conditions. A qualitative summary of the characteristics of different data dissemination protocols is presented in Table 1.3.

Table 1.3. Comparison of different self-organization protocols.

Self-Configuration Protocol	Topology	Query Specific	Fault tolerance	Energy-efficiency
Span	Flat	No	High	Low
TDMA	Flat	No	Low	Medium
Connected Sensor Cover	Flat	Yes	Medium	High
Rapid	Hierarchical	No	Low	High
Persistent	Hierarchical	No	Low	High
ASCENT	Flat	Yes	High	High

In summary, this chapter has presented an overview of wireless sensor networks, as well as efficient protocols for data dissemination and self-configuration. A qualitative comparison of these approaches has also been presented. Hierarchical cluster-based topologies appear to be well suited for sensor networks because of their fault tolerance and their ability to operate with limited resources. Self-organization protocols should be distributed, attempt to reduce the interference in inter-cluster communication, and perform data fusion. Data dissemination protocols should minimize the reception of duplicate packets, reduce the overlap of received data, and minimize the idle listening time. Energy-efficient protocols are necessary to enable the deployment of thousands of cheap sensors that are networked in an ad hoc manner. Research on this topic is therefore particularly timely and relevant.

1.7 Book Outline

The outline of the remainder of the book is as follows. Chapters 2–5 are based on the outcome of research in sensor networks carried out at Duke University. Chapter 2 addresses the coverage-driven sensor deployment problem. We first describe the virtual force algorithm and then present a set of simulation results. Next we describe the problem of uncertainty-aware sensor deployment and present two algorithms for the placement of sensors in the context of the uncertainty underlying sensor locations. Chapter 3 addresses the problem of energy-efficient target localization in cluster-based sensor networks. A two-step communication protocol is presented to reduce energy, decrease latency, and filter out false alarms. Chapter 4 shows how self-organization for energy management can be carried out in a distributed manner in sensor networks. Chapter 5 addresses energy-efficient data dissemination issues. Finally Chapters 6–7 are based on research carried out at Louisiana State University. While Chapter 6 describes energy-efficient routing, Chapter 7 is focused on time synchronization in wireless sensor networks.

2

Sensor Node Deployment

An important objective of sensor networks is to effectively monitor the environment, detect, localize, and classify targets of interest. The effectiveness of these networks is determined to a large extent by the coverage provided by the sensor deployment. The positioning of sensors affects coverage, communication cost, and resource management. In this chapter[1], we investigate two important aspects of sensor node deployment in wireless sensor networks [134, 135]. We introduce the virtual force algorithm (VFA) for sensor deployment in Section 2.2, which is a fast algorithm with low computation overhead that improves the coverage for a given number of sensors within a cluster in cluster-based sensor networks. In Section 2.4, we study the problem of uncertainty modeling for sensor node deployment in wireless sensor networks.

2.1 Sensor Node Detection Models

The sensor field is represented by a two-dimensional grid. The dimensions of the grid provide a measure of the sensor field. The granularity of the grid, i.e. distance between grid points can be adjusted to trade off computation time of the VFA algorithm with the effectiveness of the coverage measure. The detection by each sensor is modeled as a circle on the two-dimensional grid. The center of the circle denotes the sensor while the radius denotes the detection range of the sensor. We first consider a binary detection model in which a target is detected (not detected) with complete certainty by the sensor if a target is inside (outside) its circle. The binary model facilitates the understanding of the VFA model. We then investigate two types of realistic probabilistic models in which the probability that the sensor detects a target depends on the relative position of the target.

---

[1] This chapter is based on Y. Zou and K. Chakrabarty, "Sensor deployment and target localization in distributed sensor networks", *ACM Transactions on Embedded Computing Systems*, vol. 3, pp. 61-91, February 2004.

Let us consider a sensor field represented by a $m \times n$ grid. Let s be an individual sensor node on the sensor field located at grid point (x, y). Each sensor node has a detection range of r. For any grid point P at (i, j), we denote the Euclidean distance between s at (x, y) and P at (i, j) as $d_{ij}(x, y)$, i.e. $d_{ij}(x, y) = \sqrt{(x - i)^2 + (y - j)^2}$. Equation (2.1) shows the binary sensor model [23] that expresses the coverage $c_{ij}(x, y)$ of a grid point at (i, j) by sensor s at (x, y).

$$c_{ij}(x, y) = \begin{cases} 1, & \text{if } d_{ij}(x, y) < r \\ 0, & \text{otherwise.} \end{cases} \tag{2.1}$$

The binary sensor model assumes that sensor readings have no associated uncertainty. In reality, sensor detections are imprecise, hence the coverage $c_{ij}(x, y)$ needs to be expressed in probabilistic terms. A possible way of expressing this uncertainty is to assume the detection probability on a target by a sensor varies exponentially with the distance between the target and the sensor [33, 34]. This probabilistic sensor detection model given in Equation (2.2).

$$c_{ij}(x, y) = e^{-\alpha d_{ij}(x,y)} \tag{2.2}$$

This is also the coverage confidence level of this point from sensor s. The parameter α can be used to model the quality of the sensor and the rate at which its detection probability diminishes with distance. Clearly, the detection probability is 1 if the target location and the sensor location coincide. Alternatively, we can also use another probabilistic sensor detection model given in Equation (2.3), which is motivated in part by [36].

$$c_{ij}(x, y) = \begin{cases} 0, & \text{if } r + r_e \leq d_{ij}(x, y) \\ e^{-\lambda a^\beta}, & \text{if } r - r_e < d_{ij}(x, y) < r + r_e \\ 1, & \text{if } r - r_e \geq d_{ij}(x, y) \end{cases} \tag{2.3}$$

Note that $r_e(r_e < r)$ is a measure of the uncertainty in sensor detection, $a = d_{ij}(x, y) - (r - r_e)$, and λ and β are parameters that measure detection probability when a target is at distance greater than r_e but within a distance from the sensor. This model reflects the behavior of range sensing devices such as infrared and ultrasound sensors. The probabilistic sensor detection model is shown in Figure 2.1. Note that distances are measured in units of grid points. Figure 2.1 also illustrates the translation of a distance response from a sensor to the confidence level as a probability value about this sensor response. Different values of the parameters α and β yield different translations reflected by different detection probabilities, which can be viewed as the characteristics of various types of physical sensors.

It is often the case that there are obstacles in the sensor field terrain. If we are provided with such a priori knowledge about where obstacles in the sensor field, we can also build the terrain information into our models based on the principle of line of sight. An example is given in Figure 2.2. Some types of sensors are not able to see through any obstacles located in

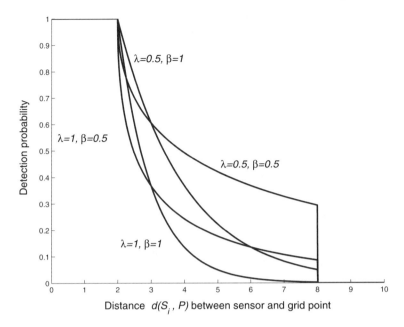

Fig. 2.1. Probabilistic sensor detection model.

the sensor field; hence models and algorithms must consider the problem of achieving an adequate sensor field coverage in presence of obstacles. Suppose C_{xy} is a $m \times n$ matrix that corresponds to the detection probabilities of each grid point in the sensor field when a sensor node is located at grid point (x, y), i.e., $C_{xy} = [c_{ij}(x, y)]_{m \times n}$. To achieve the coverage in presence of obstacles, we need to generate a mask matrix for the corresponding coverage probability matrix C_{xy} to mask out those grid points as the "blocked area", as shown in Figure 2.2. In this way, the sensor node placed at the location (x, y) will not see any grid points beyond the obstacles. We also assume that sensor nodes are not placed on any grid points with obstacles. Figure 2.3 is an example of the mask matrix for a sensor node at $(1, 1)$ in a 10 by 10 sensor field grid with obstacles located at $(7, 3), (7, 4), (3, 5), (4, 5), (5, 5)$.

2.2 Virtual Force Algorithm

As an initial sensor node deployment step, a random placement of sensors in the target area (sensor field) is often desirable, especially if no *a priori* knowledge of the terrain is available. Random deployment is also practical in military applications, where wireless sensor networks are initially established by dropping or throwing sensors into the sensor field. However, random deployment does not always lead to effective coverage, especially if the sensors are overly clustered and there is a small concentration of sensors in certain

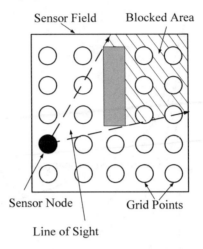

Fig. 2.2. Example to illustrate the line of sight principle.

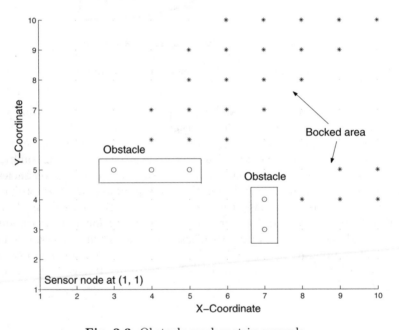

Fig. 2.3. Obstacle mask matrix example.

parts of the sensor field. However the coverage provided by a random deployment can be improved using a force-directed algorithm. We present the virtual force algorithm (VFA) as a sensor deployment strategy to enhance the coverage after an initial random placement of sensors. The VFA algorithm combines the ideas of potential field [28, 57, 79] and disk packing [78]. For a given number of sensors, VFA attempts to maximize the sensor field coverage using a combination of attractive and repulsive forces. During the execution of the force-directed VFA algorithm, sensors do not physically move but a sequence of virtual motion paths is determined for the randomly-placed sensors. Once the effective sensor positions are identified, a one-time movement is carried out to redeploy the sensors at these positions. Energy constraints are also included in the sensor repositioning algorithm. In the sensor field, each sensor behaves as a "source of force" for all other sensors. This force can be either positive (attractive) or negative (repulsive). If two sensors

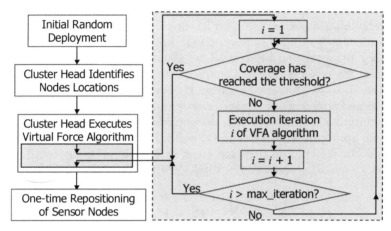

Fig. 2.4. Sensor deployment with VFA algorithm.

are placed too close to each other, the "closeness" being measured by a pre-determined threshold, they exert negative forces on each other. This ensures that the sensors are not overly clustered, leading to poor coverage in other parts of the sensor field. On the other hand, if a pair of sensors is too far apart from each (once again a pre-determined threshold is used here), they exert positive forces on each other. This ensures that a globally uniform sensor placement is achieved. Figure 2.4 illustrates how VFA algorithm is used for sensor deployment.

2.2.1 Virtual Forces

We now describe the virtual forces and virtual force calculation in the VFA algorithm. In the following discussion, we use the notation introduced in the previous subsection. Let S denote the set of deployed sensors node, i.e., $S = \{s_1, \ldots, s_k\}$ and $|S| = k$. Let the total virtual force action on a sensor node $s_p(p = 1, \ldots, k)$ be denoted by $\mathbf{F}_p$. Note that $\mathbf{F}_p$ is a vector whose orientation is determined by the vector sum of all the forces acting on s_p. Let the force exerted on s_p by another sensor $s_q(q = 1, \ldots, k, q \neq p)$ be denoted by $\mathbf{F}_{pq}$. In addition to the positive and negative forces due to other sensors, a sensor s_p is also subjected to forces exerted by obstacles and areas of preferential coverage in the grid. This provides us with a convenient method to model obstacles and the need for preferential coverage. Sensor deployment must take into account the nature of the terrain, e.g., obstacles such as building and trees in the line of sight for infrared sensors, uneven surface and elevations for hilly terrain, etc. In addition, based on relative measures of security needs and tactical importance, certain areas of the grid need to be covered with greater certainty.

The knowledge of obstacles and preferential areas implies a certain degree of *a priori* knowledge of the terrain. In practice, the knowledge of obstacles and preferential areas can be used to direct the initial random deployment of sensors, which in turn can potentially increase the efficiency of the VFA algorithm. In our virtual force model, we assume that obstacles exert repulsive (negative) forces on a sensor. Likewise, areas of preferential coverage exert attractive (positive) forces on a sensor. If more detailed information about the obstacles and preferential coverage areas is available, the parameters governing the magnitude and direction (i.e., attractive or repulsive) of these forces can be chosen appropriately. In this work, we let $\mathbf{F}_{pA}$ be the total attractive force on s_p due to preferential coverage areas, and let $\mathbf{F}_{pR}$ be the total repulsive force on s_p due to obstacles. The total force $\mathbf{F}_p$ on s_p can now be expressed as,

$$\mathbf{F}_p = \sum_{q=1,\ q\neq p}^{k} \mathbf{F}_{pq} + \mathbf{F}_{pR} + \mathbf{F}_{pA} \qquad (2.4)$$

We next express the force $\mathbf{F}_{pq}$ between s_p and s_q in polar coordinate notation. Note that $\mathbf{f} = (r, \theta)$ implies a magnitude of r and orientation θ for vector $\mathbf{f}$.

$$\mathbf{F}_{pq} = \begin{cases} (w_A(d_{pq} - d_{th}), \theta_{pq}) & \text{if } d_{pq} > d_{th} \\ 0, & \text{if } d_{pq} = d_{th} \\ (w_R\frac{1}{d_{pq}}, \theta_{pq} + \pi), & \text{if otherwise} \end{cases} \qquad (2.5)$$

where $d_{pq} = \sqrt{(x_p - x_q)^2 + (y_p - y_q)^2}$ is the Euclidean distance between sensor s_p and s_q, d_{th} is the threshold on the distance between s_p and s_q, θ_{pq} is the orientation (angle) of a line segment from s_p to s_q, and $w_A(w_R)$ is a measure of the attractive (repulsive) force. The threshold distance d_{th} controls how close sensors get to each other. As an example, consider the four sensors s_1, s_2, s_3 and s_4 in Figure 2.5. The force $\mathbf{F}_1$ on s_1 is given by $\mathbf{F}_1 = \mathbf{F}_{12} + \mathbf{F}_{13} + \mathbf{F}_{14}$. If we assume that $d_{12} > d_{th}$, $d_{13} < d_{th}$, and $d_{14} = d_{th}$, s_2 exerts an attractive force on s_1, s_3 exerts a repulsive force on s_1 and s_4 exerts no force on s_1. This is shown in Figure 2.5. Note that d_{th} is a pre-determined parameter that is supplied by the user, who can choose an appropriate value of d_{th} to achieve a desired coverage level over the sensor field.

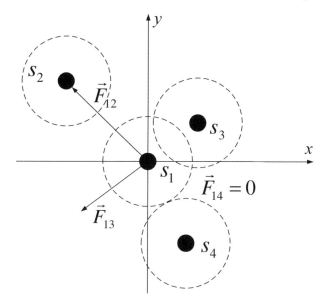

Fig. 2.5. An example of virtual forces with four sensors.

2.2.2 Overlapped Sensor Detection Areas

If $r_e \approx 0$ and we use the binary sensor detection model given by Equation (2.1), we attempt to make d_{pq} as close to $2r$ as possible. This ensures that the detection regions of two sensors do not overlap, thereby minimizing "wasted overlap" and allowing us to cover a large grid with a small number of sensors. This is illustrated in Figure 2.6(a). An obvious drawback here is that a few grid points are not covered by any sensor. Note that an alternative strategy is to allow overlap, as shown in Figure 2.6(b). While this approach ensures that all grid points are covered, it needs more sensors for grid coverage. Therefore, we adopt the first strategy. Note that in both cases, the coverage is effective only if the total area $k\pi r^2$ that can be covered with the k sensors exceeds the area of the grid.

If $r_e > 0$, r_e is not negligible and the probabilistic sensor model given by Equation (2.2) or Equation (2.3) is used. Note that due to the uncertainty in sensor detection responses, grid points are not uniformly covered with the same probability. Some grid points will have low coverage if they are covered only by only one sensor and they are far from the sensor. In this case, it is necessary to overlap sensor detection areas in order to compensate for the low detection probability of grid points that are far from a sensor. Consider a grid point with coordinate (i, j) lying in the overlap region of sensors s_p and s_q located at (x_p, y_p) and (x_q, y_q) respectively. Let $c_{ij}(s_p, s_q)$ be the probability that a target at this grid point is reported as being detected by observing the outputs of these two sensors. We assume that sensors within a cluster operate independently in their sensing activities. Thus

$$c_{ij}(s_p, s_q) = 1 - (1 - c_{ij}(s_p))(1 - c_{ij}(s_q)) \qquad (2.6)$$

where $c_{ij}(s_p) = c_{ij}(x_p, y_p)$ and $c_{ij}(s_q) = c_{ij}(x_q, y_q)$ are coverage probabilities from the probabilistic sensor detection models as we defined in Section 2.1. Since the term $1 - (1 - c_{ij}(s_p))(1 - c_{ij}(s_q))$ expresses the probability that neither s_p nor s_q covers grid point at (i, j), the probability that the grid point (i, j) is covered is given by Equation (2.6). Let c_{th} be the desired coverage threshold for all grid points. This implies that

$$\min_{i,j}\{c_{ij}(s_p, s_q)\} \geq c_{th} \qquad (2.7)$$

Note that Equation (2.6) can also be extended to a region which is overlapped by a set of k_{ov} sensors, denoted as S_{ov}, $k_{ov} = |S_{ov}|$, $S_{ov} \subseteq \{s_1, s_2, \ldots, s_k\}$. The coverage of the grid point at (i, j) due to a set of sensor nodes S_{ov} in this case is given by:

$$c_{ij}(S_{ov}) = 1 - \prod_{s_p \in S_{ov}} (1 - c_{ij}(s_p)) \qquad (2.8)$$

As shown in Equation (2.5), the threshold distance d_{th} is used to control how close sensors get to each other. When sensor detection areas overlap, the closer the sensors are to each other, the higher is the coverage probability for grid points in the overlapped areas. Note however that there is no increase in the point coverage once one of the sensors gets close enough to provide detection with a probability of one. Therefore, we need to determine d_{th} that maximizes the number of grid points in the overlapped area that satisfies $c_{ij}(s_p) > c_{th}$. Let us consider the three sensors s_1, s_2, and s_3 in Figure 2.6(a), where no overlap exists. Assume the three sensors are on a 31 by 31 grid, $r = 5$ and $r_e = 3$ in units of grid points. Figures 2.7–2.9 show how the coverage is affected by d_{th} and c_{th} when the threshold distance d_{th} is changed from $r + r_e$ to $r - r_e$. The coverage for the entire grid is calculated as the fraction of grid points that exceeds the threshold c_{th}. We can use these graphs to appropriately choose d_{th} according to the required c_{th}.

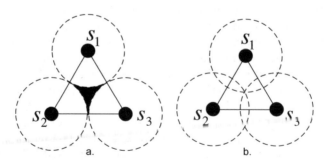

Fig. 2.6. Non-overlapped and overlapped sensor coverage areas.

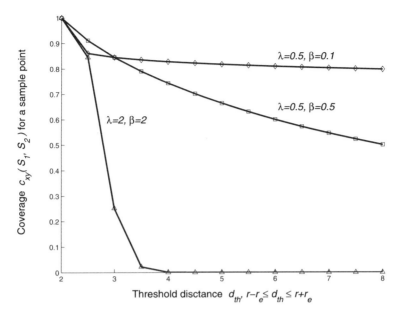

Fig. 2.7. Coverage vs. d_{th} of a sample point inside the overlapped area of s_1 and s_2.

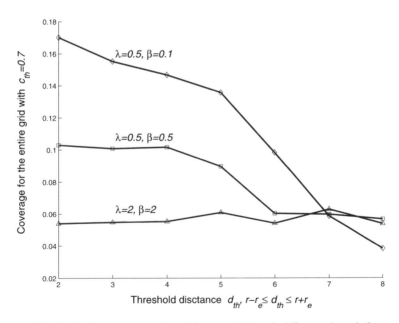

Fig. 2.8. Coverage vs. d_{th} with $c_{th} = 0.7$ and different λ and β.

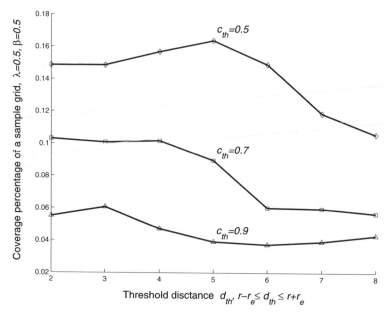

Fig. 2.9. Coverage vs. d_{th} with $\lambda = 0.5$ and $\beta = 0.5$ and $c_{th} = 0.5$, 0.7 and 0.9.

2.2.3 Energy Constraint on the VFA Algorithm

In order to prolong the battery life, the distances between the initial and final position of the sensors are limited in the repositioning phase to conserve energy. We use $d_{max}(s_p)$ to denote the maximum distance that sensor s_p can move in the repositioning phase. To simplify the discussion without loss of generality, we assume $d_{max}(s_p) = d_{max}(s_q) = d_{max}$, for $p, q = 1, 2, \ldots, k$. During the execution of the VFA algorithm, for each sensor node, whenever the distance from the current virtual position to the initial position reaches the distance limit d_{max}, any virtual forces on this sensor are disabled. For sensor s_p, let $(x_p, y_p)_{random}$ be the initial location obtained from the random deployment, and $(x_p, y_p)_{virtual}$ be the location generated by the VFA algorithm. The energy constraint can be described as:

$$\mathbf{F}_p = \begin{cases} 0, & \text{if } d((x_p, y_p)_{rand}, (x_p, y_p)_{virtual}) \geq d_{max} \\ \mathbf{F}_p, & \text{otherwise (i.e., the force is unchanged)} \end{cases} \tag{2.9}$$

Therefore the virtual force $\mathbf{F}_p$ given by Equation (2.4) on sensor s_p is ignored whenever the move violates the energy constraint expressed by d_{max}. Note that due to the energy constraint on the one-time repositioning given by Equation (2.9), it might be necessary to trade off the coverage with the energy consumed in repositioning if d_{max} is not large enough.

Note that the VFA algorithm is designed to be executed on the cluster head, which is expected to have more computational capabilities than sensor nodes. The cluster head uses the VFA algorithm to find appropriate sensor node locations based on the coverage requirements. The new locations are then sent to the sensor nodes, which perform a one-time movement to the designated positions. No movements are performed during the execution of the VFA algorithm.

2.2.4 Procedural Description

We next describe the VFA algorithm in pseudo-code. Figure 2.10 shows the data structure of the VFA algorithm and Figure 2.11 shows the implementation details in pseudo code form. For a n by m grid with a total of k sensors deployed, the computational complexity of the VFA algorithm is $O(nmk)$. Due to the granularity of the grid and the fact that the actual coverage is evaluated by the number of grid points that have been adequately covered, the convergence of the VFA algorithm is controlled by a threshold value, denoted by Δc. Let us use $c(loops)$ to denote the current grid coverage of the number $loops$ iteration in the VFA algorithm. For the binary sensor detection

---

VFA Data Structures: Grid, $\{s_1, s_2, \ldots, s_k\}$

/* n_P is the number of preferential area blocks (attractive forces) and n_O is the number of obstacle blocks (repulsive forces). S_{ij}, k_{ij} and p_table_{ij} are used for energy-aware target localization described in Chapter 3.
$(x, y)_{VFA}$ is the final position found by the VFA algorithm. d_{max} is the energy constraint on the sensor re-positioning phase in the VFA algorithm. */
1 *Grid* structure:
2   Properties: *width, height*, k, c_{th}, d_{th}, $c(loops)$, $\bar{c}$, Δc;
3   Preferential areas: $PA_i(x, y, wx, wy)$, $i = 1, 2, \ldots, n_P$;
4   Obstacles areas: $OA_i(x, y, wx, wy)$, $i = 1, 2, \ldots, n_O$;
5   Grid points, P_{ij}: $c_{ij}(\{s_1, s_2, \ldots, s_k\})$, $S_{ij}, k_{ij}, p\_table_{ij}$;
6 Sensor s_p structure:
    $(x_p, y_p)_{random}, (x_p, y_p)_{virtual}, (x, y)_{VFA}, p, r, r_e, \alpha, \beta, d_{max}$;

Fig. 2.10. Data structures used in the VFA algorithm.

---

model without the energy constraint, the upper bound value denoted as $\bar{c}$ is $k\pi r^2$; for the probabilistic sensor detection model or binary sensor detection model with the energy constraint, $c(loops)$ is checked for saturation by defining $\bar{c}$ as the average of the coverage ratios of the near 5 (or 10) iterations. Therefore, the VFA algorithm continues to iterate until $|c(loops) - \bar{c}| \le \Delta c$. In our experiments, Δc is set to 0.001.

Note that there exists the possibility of certain pathological scenarios in which the VFA algorithm is rendered ineffective, e.g., if the sensors are initially placed along the circumference of a circle such that all virtual forces are

---

Procedure *Virtual_Force_Algorithm* (Grid, $\{s_1, s_2, \ldots, s_k\}$)

---

1 Set $loops = 0$;
2 Set $MaxLoops =$**MAX_LOOPS**;
3 **While** ($loops < MaxLoops$)
4   /* coverage evaluation */
5   **For** grid point P at (i,j) in Grid, $i \in [1, width], j \in [1, height]$
6   **For** $s_p \in \{s_1, s_2, \ldots, s_k\}$
7   Calculate $c_{ij}(x_p, y_p)$ from the sensor model using $(d_{ij}(x_p, y_p), c_{th}, d_{th}, \alpha, \beta)$;
8   **End**
9   **End**
10   **If** coverage requirements are met: $|c(loops) - \bar{c}| \leq \Delta c$
11   **Break** from **While** loop;
12   **End**
13   /* virtual forces among sensors */
14   **For** $s_p \in \{s_1, s_2, \ldots, s_k\}$
15   Calculate $\mathbf{F}_{pq}$ using $d(s_p, s_q)$, d_{th}, w_A, w_R;
16   Calculate $\mathbf{F}_{pA}$ using $d(s_p, PA_1, \ldots, PA_{n_P})$, d_{th};
17   Calculate $\mathbf{F}_{pR}$ using $d(s_p, OA_1, \ldots, OA_{n_O})$, d_{th};
18        $\mathbf{F}_p = \sum \mathbf{F}_{pq} + \mathbf{F}_{pR} + \mathbf{F}_{pA}, q = 1, \ldots, k, q \neq p$;
19   **End**
20   /*move sensors virtually */
21   **For** $s_p \in \{s_1, s_2, \ldots, s_k\}$
22   /* energy constraint on the sensor movement */
23   **If** $d((x_p, y_p)_{random}, (x, y)_{virtual}) \geq d_{max}$
24   Set $\mathbf{F}_p = 0$;
25   **End**
26        $\mathbf{F}_p$ virtually moves s_p to its next position;
27   **End**
28   /* continue to next iteration */
29   Set $loops = loops + 1$;
30 **End**

Fig. 2.11. Pseudocode of the VFA algorithm.

balanced. The efficiency of the VFA algorithm depends on the values of the force parameters w_A and w_R. We found that the algorithm converged more rapidly for our case studies if $w_R \gg w_A$. This need not always be true, so we are examining ways to choose appropriate values for w_R and w_A based on the initial configuration.

2.3 VFA Simulation Results

In this section, we present simulation results obtained using the VFA algorithm. The deployment requirements include the maximum improvement of coverage over random deployment, the coverage for preferential areas and the avoidance of obstacles. For all simulation results presented in this section, distances are measured in units of grid points. A total of 20 sensors are placed in the sensor field in the random placement stage. Each sensor has a detection radius of 5 units ($r = 5$), and range detection error of 3 units ($r_e = 3$) for the probabilistic detection model. The sensor field is 50 by 50 in dimension. The simulation is done on a Pentium III 1.0GHz PC using Matlab.

2.3.1 Case Study 1

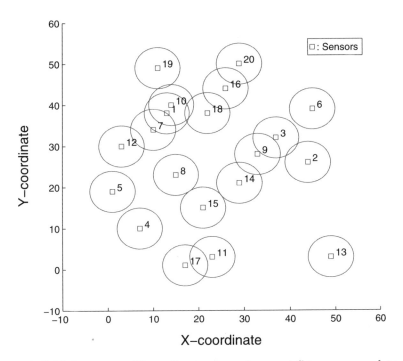

Fig. 2.12. Initial sensor positions after random placement (binary sensor detection model).

Figures 2.12–2.15 present simulation results based on the binary sensor detection model given by Equation (2.1). The initial locations of the sensors are shown in Figure 2.12. Figure 2.13 shows the final sensor positions determined by the VFA algorithm. For the binary sensor detection model, an

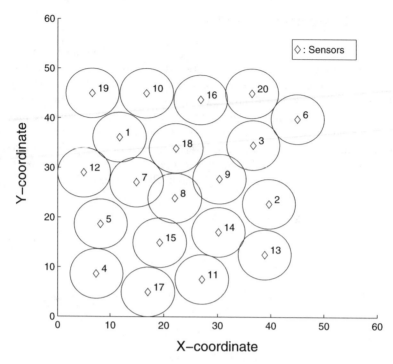

Fig. 2.13. Sensor positions after the execution of the VFA algorithm (binary sensor detection model).

upper bound on the coverage is given by the ratio of the sum of the circle areas (corresponding to sensors) to the total area of the sensor field. For our example, this upper bound evaluates to 0.628 and it is achieved after 28 iterations of the VFA algorithm. Figure 2.14 shows the virtual movement traces of all sensors during the execution of the VFA algorithm. Figure 2.15 shows the improvement in coverage during the execution of the VFA algorithm.

2.3.2 Case Study 2

Figures 2.16–2.18 present simulation results for the probabilistic sensor model given by Equation (2.3). The probabilistic sensor detection model parameters are set as $\lambda = 0.5$, $\beta = 0.5$, and $c_{th} = 0.7$. The initial sensor placements are shown in Figure 2.16. Figure 2.17 shows the final sensor positions determined by the VFA algorithm. Figure 2.18 shows the virtual movement traces of all sensors during the execution of the VFA algorithm. We can see that overlap areas are used to increase the number of grid points whose coverage exceeds the required threshold c_{th}. Figure 2.19 shows the improvement of coverage during the execution of the VFA algorithm. Note that the upper bound for the coverage for the probabilistic sensor detection model in Figure 2.19

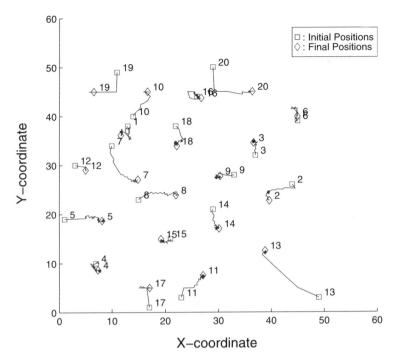

Fig. 2.14. A trace of virtual moves made by the sensors (binary sensor detection model).

(roughly 0.38) is lower than the upper bound for the case of binary sensor detection model in Figure 2.15 (roughly 0.63). This due to the fact that for the simulation results shown here, the coverage for the binary sensor detection model is the fraction of the sensor field covered by the circles. For the probabilistic sensor detection model, even though there are a large number of grid points that are covered, the overall number of grid points with coverage probability greater than the required level is fewer.

2.3.3 Case Study 3

As discussed in Section 2.2, VFA is also applicable to a sensor field containing obstacles and preferential areas. If obstacles are to be avoided, they can be modeled as repulsive force sources in the VFA algorithm. Preferential areas should be covered first, therefore they are modeled as attractive force sources in the VFA algorithm. Figure 2.20–2.23 present simulation results for a 50 by 50 sensor field that contains an obstacle and a preferential area. The binary sensor detection mobel given by Equation (2.1) is used for this simulation. The initial sensor placements are shown in Figure 2.20. Figure 2.21 shows the final sensor positions determined by the VFA algorithm. Figure 2.22 shows

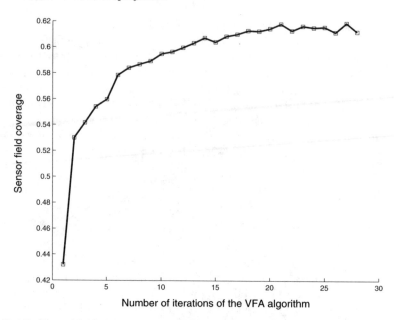

Fig. 2.15. Sensor field coverage improvement by the VFA algorithm (binary sensor detection model).

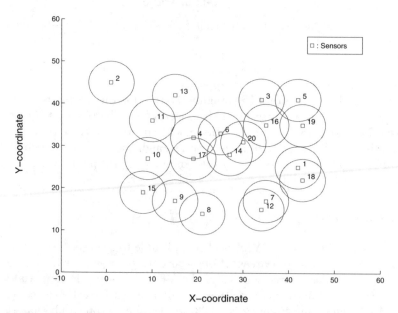

Fig. 2.16. Initial sensor positions after random placement (probabilistic sensor detection model).

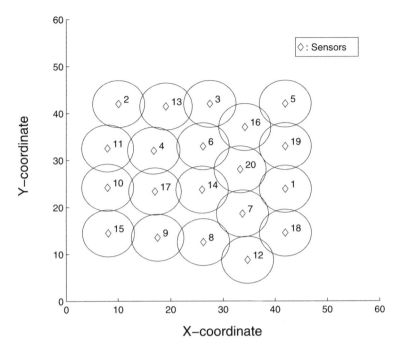

Fig. 2.17. Sensor positions after the execution of the VFA algorithm (probabilistic sensor detection model).

the virtual movement traces of all sensors during the execution of the VFA algorithm. Figure 2.23 shows the improvement of coverage during the execution of the VFA algorithm.

The VFA algorithm does not require much computation time. For Case study 1, the VFA algorithm took only 25 seconds for 30 iterations. For Case study 2, the VFA algorithm took only 3 minutes to complete 50 iterations. Finally for Case study 3, the VFA algorithm took only 48 seconds to complete 50 iterations. Note that these computation times include the time needed for displaying the simulation results on the screen. CPU time is important because sensor redeployment should not take excessive time.

In order to examine how the VFA algorithm scales for larger problem instances, we considered up to 90 sensor nodes in a cluster for a 50 by 50 grid, with $r = 3$, $r_e = 2$, $\lambda = 0.5$ and $\beta = 0.5$ for all cases. For a given number of sensor nodes, we run the VFA algorithm over 10 sets of random deployment results and take the average of the computation time. The results, listed in Table 2.1, show that the CPU time grows slowly with the number of sensors k. For a total of 90 sensors, the CPU time is only 4 minutes on a Pentium III PC. In practice, a cluster head usually has less computational power than a Pentium III PC; however, our results indicate that even if the cluster head

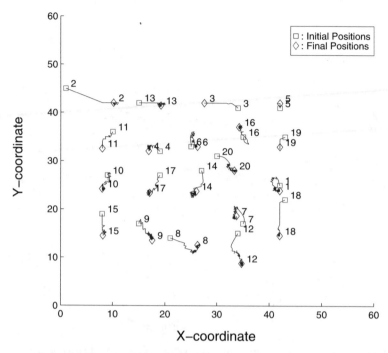

Fig. 2.18. A trace of virtual moves made by the sensors (probabilistic sensor detection model).

has less memory and an on-board processor that runs 10 times slower, the CPU time for the VFA algorithm is reasonable.

Table 2.1. The computation time for the VFA algorithm for larger problem instances.

k	Binary Model	Probabilistic Model	k	Binary Model	Probabilistic Model
40	21 seconds	1.8 minutes	70	46 seconds	3.6 minutes
50	32 seconds	2.2 minutes	80	59 seconds	3.7 minutes
60	38 seconds	3.1 minutes	90	64 seconds	4.0 minutes

2.4 Uncertainty Modeling

The topology of the sensor field, i.e., the locations of the sensors, determines to a large extent the quality and the extent of the coverage provided by the sensor network. However, even if the sensor locations are precomputed for

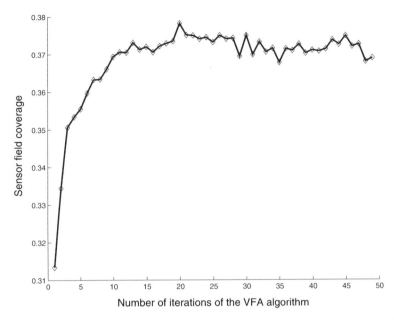

Fig. 2.19. Sensor field coverage achieved using the VFA algorithm (probabilistic sensor detection model).

optimal coverage and resource utilization, there are inherent uncertainties in the sensor locations when the sensors are dispersed, scattered, or airdropped. Thus a key challenge in sensor deployment is to determine an uncertainty-aware sensor field architecture that reduces cost and provides high coverage, even though the exact location of the sensors may not be controllable. We consider the sensor deployment problem in the context of uncertainty in sensor locations subsequent to airdropping[138, 139]. Sensor deployment in such scenarios is inherently non-deterministic and there is a certain degree of randomness associated with the location of a sensor in the sensor field. We present two algorithms for the efficient placement of sensors in a sensor field when the exact locations of the sensors are not known. The proposed approach is aimed at optimizing the number of sensors and determining their placement to support distributed sensor networks. These algorithms are targeted at average coverage as well as at maximizing the coverage of the most vulnerable regions in the sensor field. Experimental results for an example sensor field demonstrate the application of our approach.

In applications such as battlefield surveillance and environmental monitoring, sensors may be dropped from airplanes. Such sensors cannot be expected to fall exactly at predetermined locations; rather there are regions where there is a high probability of sensor being actually located (Figure 2.24). In underwater deployment, sensors may move due to drift or water currents. Furthermore in most real-life situations, it is difficult to pinpoint the exact location

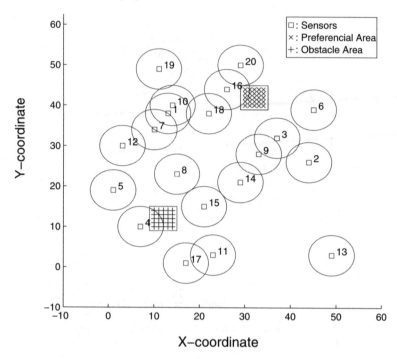

Fig. 2.20. Initial sensor positions after random placement with obstacles and preferred areas.

of each sensor since only a few of the sensors may be aware of their locations. Thus the position of sensors may not exactly known and for every point in the sensor field, there is only a certain probability of a sensor being located at that point.

In this section, we present two algorithms for sensor deployment wherein we assumed that sensor positions are not exactly predetermined. We assume that the sensor locations are calculated before deployment and an attempt is made during the airdrop to place sensors at these locations; however, the sensor placement calculations and coverage optimization are based on a Gaussian model, which assumes that if a sensor is intended for a specific point P in the sensor field, its exact location can be anywhere in a "cloud" surrounding P.

2.4.1 Modeling of Non-Deterministic Placement

During sensor deployment, an attempt is made to place sensors at appropriate predetermined locations by air-dropping or other means. This does not guarantee however that sensors are actually placed at the designated positions, due to unanticipated conditions such as wind, the slope of the terrain, etc. In this case, there is a certain probability of a sensor being located at a particular grid point as a function of the designated location. The deviation about

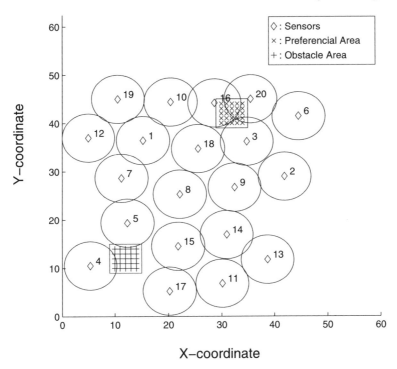

Fig. 2.21. Sensor positions after the execution of the VFA algorithm with obstacles and preferred areas.

the designated sensor locations may be modeled using a Gaussian probability distribution, where the intended coordinates (x, y) serve as the mean values with standard deviation σ_x and σ_y in the x and y dimensions, respectively. Assuming that the deviations in the x and y dimensions are independent, the joint probability density function with mean (x, y) is given by:

$$p_{xy}(x', y') = \frac{e^{-\frac{(x'-x)^2}{2\sigma_x^2} - \frac{(y'-y)^2}{2\sigma_y^2}}}{2\pi\sigma_x\sigma_y}. \tag{2.10}$$

Let us use the notation introduced in the previous section. We still consider a sensor field represented by a $m \times n$ grid, denoted as $Grid$, with S denoting the set of sensor nodes. Let L_S be the set that contains corresponding sensor node locations, i.e. $L_S = \{(x_p, y_p)|s_p$ at $(x_p, y_p), s_p \in S\}$. Let A be the total area encompassing all possible sensor locations. To model the uncertainty in sensor locations, the conditional probability $c_{ij}^*(x, y)$ for a grid point (i, j) to be detected by a sensor that is supposed to be deployed at (x, y) is then given by:

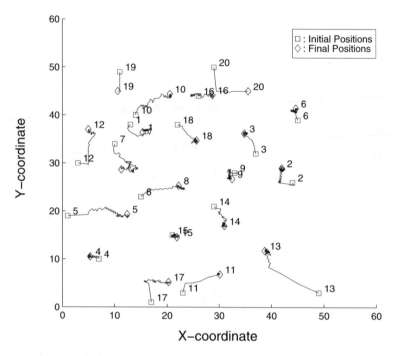

Fig. 2.22. A trace of virtual moves made by the sensors with obstacles and preferred areas.

$$c_{ij}^*(x, y) = \frac{\displaystyle\sum_{(x',y')\in A} c_{ij}(x', y')p_{xy}(x', y')}{\displaystyle\sum_{(x',y')\in A} p_{xy}(x', y')}. \tag{2.11}$$

Based on Equations (2.10) and (2.11), we defined the matrices $C_{xy}^* = [c_{ij}^*(x, y)]_{m \times n}$ and $P = [p_{xy}(x', y')]_A$. Figure 2.25 illustrates the effect of the uncertainty under different variations in the grid point coverage probability $c_{ij}(x, y)$ due to a sample sensor at the designated placement location of $(x, y) = (3, 4)$ for a 10 by 10 grid.

2.4.2 Uncertainty-Aware Placement Algorithms

In this section, we introduce the sensor placement algorithm with consideration of uncertainties in sensor locations. The goal of sensor placement algorithms is to determine the minimum number of sensors and their locations such that every grid point is covered with a minimum confidence level. The sensor placement algorithms do not give us the actual location of the sensor but only the mean position of the sensor. It is straightforward to define the miss probability in our sensor deployment scenario. The *miss probability* of a

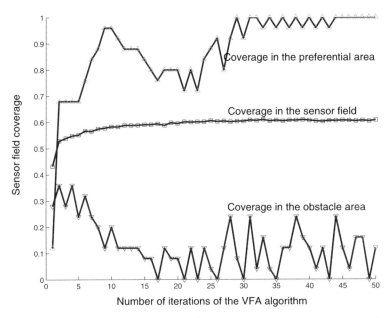

Fig. 2.23. Sensor field coverage achieved using the VFA algorithm with obstacles and preferred areas.

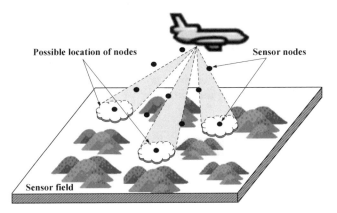

Fig. 2.24. Sensors dropped from airplanes. The clouded region gives the possible region of a sensor location. The black dots within the clouds show the mean (intended) position of a sensor.

grid point (i, j) due to a sensor at (x, y), denoted as $m_{ij}(x, y)$, is given by:

$$m_{ij}(x, y) = 1 - c_{ij}^*(x, y). \tag{2.12}$$

Therefore the miss probability matrix due to a sensor placed at (x, y) is $M_{xy} = [m_{ij}(x, y)]_{m \times n}$. M_{xy} is associated with each grid point and can be pre-determined based on Equations (2.10), (2.11) and (2.12). Since a number

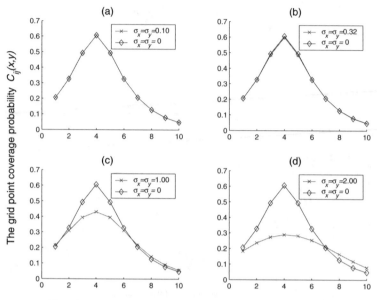

Gid points $1 \leq i \leq 10$, $j=2$ with a sample sensor at designated of (x=3, y=4)

Fig. 2.25. Example of grid point coverage under uncertainty for (a) $\sigma_x = \sigma_y = 0.10$ (b) $\sigma_x = \sigma_y = 0.32$ (c) $\sigma_x = \sigma_y = 1.00$ (d) $\sigma_x = \sigma_y = 2.00$.

of sensors are placed for coverage, we would like to know the miss probability of each grid point due to a set of sensors, namely the collective miss probability. We denote the term *collective miss probability* as m_{ij} and define it in the form of maximum likelihood function as

$$m_{ij} = \prod_{(x,y) \in L_S} m_{ij}(x, y) = \prod_{(x,y) \in L_S} [1 - c_{ij}^*(x, y)]. \qquad (2.13)$$

Accordingly we have $M = [m_{ij}]_{m \times n}$ as the collective miss probability matrix over the grid points in the sensor field.

We determine the location of the sensors one at a time. In each step, we find all possible locations that are available on the grid for a sensor, and calculate the overall miss probability associated due to this sensor and those already deployed. We denote the *overall miss probability* due to the newly introduced sensor at grid point (x, y) as $\widetilde{m}(x, y)$, which is defined as

$$\widetilde{m}(x, y) = \sum_{(i,j) \in Grid} m_{ij}(x, y) m_{ij}. \qquad (2.14)$$

Based on the $\widetilde{m}(x, y)$ values, where $(x, y) \in Grid$ and $(x, y) \notin L_S$, we can place sensors either at the grid point with the maximum miss probability (the worst coverage case) or the minimum miss probability (the best coverage case).

We refer to the two strategies as MAX_MISS and MIN_MISS, respectively. Therefore, the sensor location can be found based on the following rule. For $(x, y) \in Grid$ and $(x, y) \notin L_S$,

$$\widetilde{m}(x, y) = \begin{cases} \min\{\widetilde{m}(x', y')\}, & \text{if MIN\_MISS is used;} \\ \max\{\widetilde{m}(x', y')\}, & \text{if MAX\_MISS is used.} \end{cases} \tag{2.15}$$

When the best location is found for the current sensor, the collective miss probability matrix M is updated with the newly introduced sensor at location (x, y). This is carried out using Equation (2.16):

$$M = M \cdot M_{xy} = [m_{ij} \cdot m_{ij}(x, y)]_{m \times n}. \tag{2.16}$$

There are two parameters that serve as the termination criterion for the two algorithm. The first is k_{max}, which is the maximum number of sensors that we can afford to deploy. The second is the threshold on the miss probability of each grid point, m_{th}. Our objective is to ensure that every grid point is covered with probability at least $c_{th} = 1 - m_{th}$. Therefore, the rule to stop the further execution of the algorithm is

$$m_{ij} < m_{th} \text{ for all } (i, j) \in Grid \text{ or } k > k_{max}, \tag{2.17}$$

where k is the number of deployed sensors. The performance of the proposed algorithm is evaluated using the *average coverage probability* of the grid defined as

$$c_{avg} = \frac{\displaystyle\sum_{(x,y) \in Grid} c_{ij}}{m \cdot n}. \tag{2.18}$$

where c_{ij} is the *collective coverage probability* of a grid point due to all sensors on the grid, defined as

$$c_{ij} = 1 - \prod_{(x,y) \in L_S} m_{ij}(x, y)$$

$$= 1 - \{ \prod_{(x,y) \in L_S} [1 - c_{ij}^*(x, y)] \}. \tag{2.19}$$

We have thus far considered the coverage of only the grid points in the sensor field. In order to provide robust coverage of the sensor field, we also need to ensure that the region that lies between the grid points is adequately covered, i.e., every non-grid point has a miss probability less than the threshold m_{th}. Consider the four grid points in Figure 2.26 that lie on the four corners of a square. Let the distance between these grid points be d^*. The point of intersection of the diagonals of the square is at distance $\frac{d^*}{\sqrt{2}}$ from the four grid points. The following theorem provides a sufficient condition under which the non-grid points are adequately covered by the MIN_MISS and MAX_MISS algorithms.

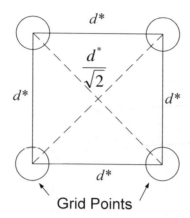

Fig. 2.26. Coverage of non-grid points.

Theorem 2.1. *Let the distance between the grid point P_1 and a potential sensor location P_2 be d. Let the distance between adjacent grid points be d^*. If a value of $d + \frac{d^*}{\sqrt{2}}$ is used to calculate the coverage of grid point P_1 due to a sensor at P_2, and the number of available sensors is adequate, the miss probability of all the non-grid points is less than a given threshold m_{th} when the algorithms MAX_MISS and MIN_MISS terminate.*

Proof: Consider the four grid points in Figure 2.26. The center of square, i.e., the point of intersection of diagonals, is at a distance of $\frac{d^*}{\sqrt{2}}$ from each of the four grid points. Every other non-grid point is at a shorter distance (less than $\frac{d^*}{\sqrt{2}}$) from at least one of the four grid points. Thus if a value of $d + \frac{d^*}{\sqrt{2}}$ is used to determine coverage in the MAX_MISS and MIN_MISS algorithms, we can guarantee that every non-grid point is covered with a probability that exceeds $1 - m_{th}$. ∎

In order to illustrate Theorem 1, we consider a 5 by 5 grid with $\alpha = 0.5, \lambda = 0.5, \beta = 0.5$ and $m_{th} = 0.4$. We use Theorem 1 and the MAX_MISS algorithm to determine sensor placement and to calculate the miss probabilities for all the centers of the squares. The results are shown in Figure 2.27 and Figure 2.28 for both sensor detection models. They indicate that the miss probabilities are always less than the threshold m_{th}, thereby ensuring adequate coverage of the non-grid points.

2.4.3 Procedural Description

Note that matrices C_{xy}, M_{xy} and P_A can all be calculated before the actual execution of the placement algorithms. This is illustrated in Figure 2.29 as the pseudocode for the initialization procedure. The initialization procedure is the algorithm overhead which has a complexity of $O((mn)^2)$, where the dimension of the grid is $m \times n$. Once the initialization is done, we may apply either

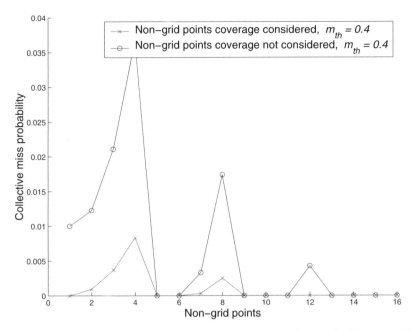

Fig. 2.27. Coverage of non-grid points for the sensor model given by Equation (2.2).

MIN_MISS or MIN_MISS uncertainty-aware sensor placement algorithm using different values for m_{th} and k_{max} with the same C_{xy}, M_{xy} and P_A. Figure 2.30 outlines the main part in pseudocode for the uncertainty-aware sensor placement algorithms. The computational complexity for both MIN_MISS and MAX_MISS is $O(mn)$.

2.5 Simulation Results

Next we present simulation results for the proposed uncertainty-aware sensor placement algorithms MIN_MISS and MAX_MISS using the same testing platform. Note that for practical reasons, we use a truncated Gaussian model because the sensor deviations in a sensor location are unlikely to span the complete sensor field. Therefore $x' - x$ and $y' - y$ in Equation (2.10) are limited to a certain range, which reflects how large the variation is in the sensor locations during the deployment. The maximum error in x direction is denoted as $e^x_{max} = \max(x' - x)$, and the maximum error in y direction is denoted as $e^y_{max} = \max(y' - y)$. We then present our simulation results for different sets of parameters in units of grid point where $m = n = 10$, $\sigma_x = \sigma_y = 0.1, 0.32, 1, 2$, and $e^x_{max} = e^y_{max} = 2, 3, 5$.

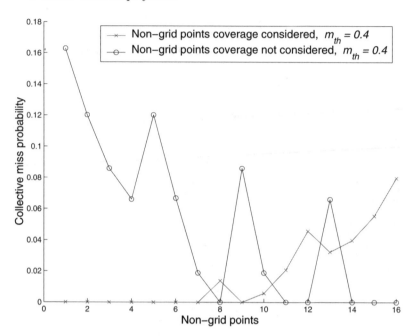

Fig. 2.28. Coverage of non-grid points for the sensor model given by Equation (2.3).

2.5.1 Case Study 1

We first consider the probabilistic sensor detection model given by Equation (2.2) with $\alpha = 0.5$. Figure 2.31 presents the result for the two sensor placement algorithms described by Equation (2.15). Figure 2.32 compares the proposed MIN_MISS and MAX_MISS algorithms with the base case where no location errors are considered, i.e. an uncertainty-oblivious (UO) strategy is followed by setting $\sigma_x = \sigma_y = 0$. We also consider a random deployment of sensors. The results show that MIN_MISS is nearly as efficient as the base uncertainty-oblivious algorithm, yet it is much more robust. Figure 2.33 presents results for the truncated Gaussian models with different maximum errors. Compared to random deployment, MIN_MISS requires more sensors here but we expect random deployment to perform worse in the presence of obstacles. Figure 2.34 for MIN_MISS and MAX_MISS with coverage obtained without location uncertainty. The results show that the MAX_MISS algorithm, which place more sensors for a given coverage threshold, provides higher overall coverage.

2.5.2 Case Study 2

Next, we consider the probabilistic sensor detection model given by Equation (2.3) with $r = 5, r_e = 4, \lambda = 0.5$, and $\beta = 0.5$. Figure 2.35 presents the result for the two sensor placement algorithms described by Equation (2.15). Figure 2.36 compares the proposed MIN_MISS and MAX_MISS algorithms with

---

Procedure *NDSP_Proc_Init* (`Grid`, $\sigma_x, \sigma_y, \alpha, \lambda, \beta$)

---

01 /* Build the uncertainty area matrix $P = [p_{xy}(x', y')]_A$ */
02 **For** $(x', y') \in A$

03 $\qquad p_{xy}(x', y') = \dfrac{e^{-\frac{(x'-x)^2}{2\sigma_x^2} - \frac{(y'-y)^2}{2\sigma_y^2}}}{2\pi\sigma_x\sigma_y}$

04 **End**
05 /* Build the miss probability matrix for all grid points. */
06 **For** grid point $(x, y) \in Grid$
07 $\qquad$ /* Build C_{xy}, C_{xy}^*, and M_{xy} for sensor node at (x, y). */
08 $\qquad$ **For** grid point $(i, j) \in Grid$
09 $\qquad\qquad$ /* Non-grid points coverage are considered based on Theorem 1. */
09 $\qquad\qquad d_{ij}(x, y) = \sqrt{(x-i)^2 + (y-j)^2} + \frac{d^*}{\sqrt{2}}$;
10 $\qquad\qquad$ /* Calculate the grid point coverage probability based on the
sensor detection model. */
11 $\qquad\qquad$ Calculate $c_{ij}(x, y)$:
12 $\qquad\qquad$ /* Sensor detection model 1. */
13 $\qquad\qquad$ Model 1: $c_{ij}(x, y) = e^{-\alpha d_{ij}(x,y)}$
14 $\qquad\qquad$ /* Sensor detection model 2. */

15 $\qquad\qquad$ Model 2: $c_{ij}(x, y) = \begin{cases} 0, & \text{if } r + r_e \leq d_{ij}(x, y) \\ e^{-\lambda a^\beta}, & \text{if } |r - d_{ij}(x, y)| < r_e \\ 1, & \text{if } r - r_e \geq d_{ij}(x, y) \end{cases}$

16 $\qquad\qquad$ /* Modeling of uncertainty in sensor node locations. */

17 $\qquad\qquad c_{ij}^*(x, y) = \dfrac{\sum\limits_{(x',y')\in A} c_{ij}(x', y')p_{xy}(x', y')}{\sum\limits_{(x',y')\in A} p_{xy}(x', y')}$;

18 $\qquad\qquad$ /* The miss probability matrix */
19 $\qquad\qquad m_{ij}(x, y) = 1 - c_{ij}^*(x, y)$;
20 $\qquad$ **End**
21 $\qquad$ /* Use the obstacle mask matrix based on the a priori knowledge about
the terrain. */
22 $\qquad$ **If** Obstacles exist
23 $\qquad\qquad C_{xy} = C_{xy} \cdot ObstacleMaskMatrix$
24 $\qquad\qquad$ Revise M_{xy}.
25 $\qquad$ **End**
26 **End**
27 /* Initially the overall miss probability matrix is set to I. */
28 $M = [m_{ij}]_{m \times n} = [1]_{m \times n}$;

Fig. 2.29. Initialization pseudocode.

the base case where no location errors are considered. Figure 2.37 presents
results for the truncated Gaussian models with different maximum errors.

Procedure $NDSP\_Proc\_Main$ (**type**, k_{max}, m_{th}, **Grid**, C_{xy}, M_{xy}, P_A, M)

01 /* Initially no sensors have been placed yet. */
02 Set $S = \phi$; $L_S = \{\phi\}$; $k = |S|$;
03 /* Repeatedly placing sensors until requirement is satisfied.*/
04 **Rep eat**
05   /* Evaluate the miss probability due to a sensor at (x, y). */
06   **For** grid point $(x, y) \in Grid$ **And** $(x, y) \notin L_S$
07   Retrieve $M_{xy} = [m_{ij}(x, y)]_{m \times n}$
                $= [1 - c_{ij}^*(x, y)]_{m \times n}$

$$= \left[1 - \frac{\displaystyle\sum_{(x', y') \in A} c_{ij}(x', y') p_{xy}(x', y')}{\displaystyle\sum_{(x', y') \in A} p_{xy}(x', y')} \right]_{m \times n}$$

08   /* Miss probability if sensor node is placed at (x, y) */
09          $\widetilde{m}(x, y) = \displaystyle\sum_{(i, j) \in Grid} m_{ij}(x, y) m_{ij}$;
10   **End**
11   /* Place sensor node using selected algorithm. */
12   **If** $type$ =MIN_MISS
13   Find $(x, y) \in Grid$ and $(x, y) \notin L_s$ such that
$\widetilde{m}(x, y) = \min\{\widetilde{m}(x', y')\}$, $(x', y') \in Grid$;
14   **Else** /* MAX_MISS */
15   Find $(x, y) \in Grid$ and $(x, y) \notin L_s$ such that
$\widetilde{m}(x, y) = \max\{\widetilde{m}(x', y')\}$, $(x', y') \in Grid$;
16   **End**
17   /* Save the information of sensor node just placed. */
18   Set $k = k + 1$;
19   Set $L_S = L_S \cup \{(x, y)\}$;
20   Set $S = S \cup \{s_k\}$;
20   /* Update current overall miss probability matrix. */
21   **For** grid point $(i, j) \in Grid$
22          $m_{ij} = m_{ij} \cdot m_{ij}(x, y)$;
23   **End**
24 /* Check if the placement requirement is satisfied. */
25 **Until** $m_{ij} < m_{th}$ for all $(i, j) \in Grid$ **Or** $k > k_{max}$;

Fig. 2.30. Pseudocode for sensor placement algorithm.

Figure 2.38 compares the coverage based on Equation (2.18) for MIN_MISS and MAX_MISS with coverage obtained without location uncertainty. We notice that due to the different probability values as a reflection of the confidence level in sensor responses from these two different models, the results in sensor placement are also different. Compared with Case study 1, this sensor detec-

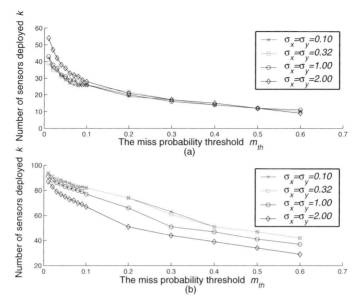

Fig. 2.31. Number of sensors required as a function of the miss probability threshold with $\alpha = 0.5, e^x_{max} = e^y_{max} = 5$, for (a) MIN_MISS (b) MAX_MISS.

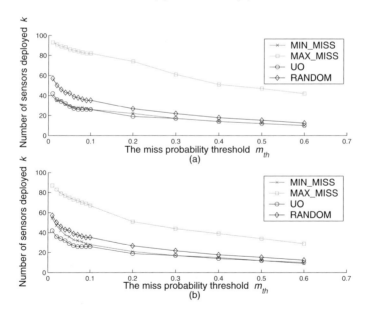

Fig. 2.32. Number of sensors required for various placement schemes with $\alpha = 0.5, e^x_{max} = e^y_{max} = 5$, and (a) $\sigma_x = \sigma_y = 0.32$ (b) $\sigma_x = \sigma_y = 2$.

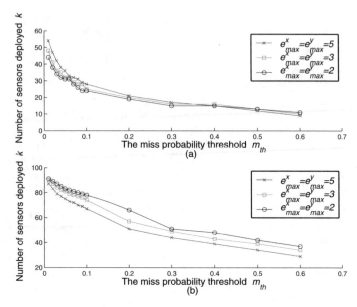

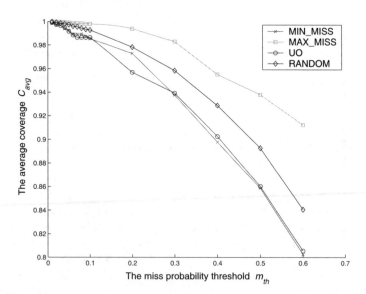

Fig. 2.33. Comparisons in different truncated Gaussian models with $\alpha = 0.5, \sigma_x = \sigma_y = 2$ for (a) MIN_MISS (b) MAX_MISS.

Fig. 2.34. Comparison in average coverage for various placement schemes with $\alpha = 0.5, e^x_{max} = e^y_{max} = 5, \sigma_x = \sigma_y = 0.32$.

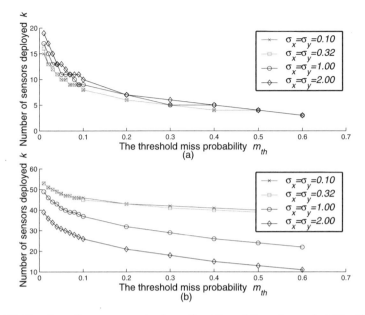

Fig. 2.35. Number of sensors required as a function of the miss probability threshold with $\alpha = 0.5, e_{max}^x = e_{max}^y = 5$, for (a) MIN_MISS (b) MAX_MISS.

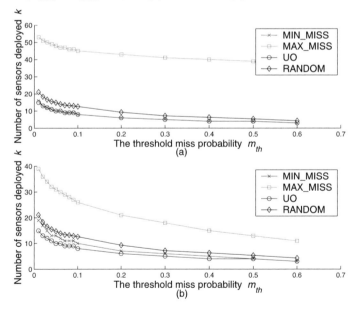

Fig. 2.36. Number of sensors required for various placement schemes with $\alpha = 0.5, e_{max}^x = e_{max}^y = 5$, and (a) $\sigma_x = \sigma_y = 0.32$ (b) $\sigma_x = \sigma_y = 2$.

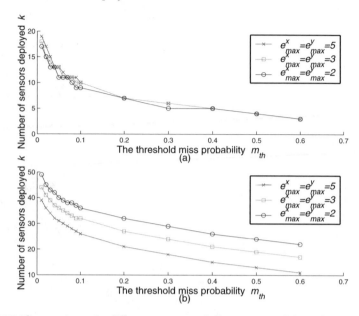

Fig. 2.37. Comparisons in different truncated Gaussian models with $\alpha = 0.5, \sigma_x = \sigma_y = 2$ for (a) MIN_MISS (b) MAX_MISS.

tion model with the selected model parameters as $\lambda = 0.5$ and $\beta = 0.5$ requires less number of sensor nodes for the same miss probability threshold. Part of the reason is due to the fact that in Equation (2.3), we have full confidence in sensor responses for grid points that are very close to the sensor node, i.e. $c_{ij}(x,y) = 1$ if $r - r_e \geq d_{ij}(x,y)$. However, this case study shows that the proposed sensor deployment algorithms do not depend on any specific type of sensor models. The sensor detection model can be viewed as a plug-in module when different types of sensors are encountered in applying the deployment algorithms.

2.5.3 Case Study 3

Next we consider a terrain model with the existence of obstacles. We have manually placed one obstacle that occupies grid points $(7,3), (7,4)$, and another obstacle that occupies grid points $(3,5), (4,5), (5,5)$. They are marked as "Obstacle" in Figure 2.3, which gives the layout of the setup for this case study. We have evaluated the proposed algorithms on the sensor detection model in Case study 2, which is given by Equation (2.3) with the same model parameters as $r = 5, r_e = 4, \lambda = 0.5$, and $\beta = 0.5$. Figure 2.39 presents results for the truncated Gaussian models with different maximum errors. Figure 2.40 compares the coverage based on Equation (2.18) for MIN_MISS and MAX_MISS with coverage obtained without location uncertainty. It is obvious that because of the existence of obstacles, the actual range of sensor

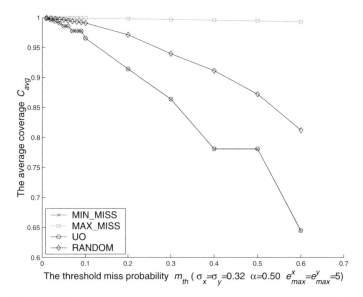

Fig. 2.38. Comparison in average coverage for various placement schemes with $\alpha = 0.5, e^x_{max} = e^y_{max} = 5, \sigma_x = \sigma_y = 0.32$.

detection due to the line-of-sight principle. Therefore, the reduction in sensor detection range causes an increase in the number of sensors required for the same miss probability threshold, as shown in Figure 2.39 and Figure 2.40.

2.6 Discussion

In this chapter, we have discussed two important aspects in sensor node deployment for wireless sensor networks. The proposed VFA algorithm introduced in Section 2.2 improves the sensor field coverage considerably compared to random sensor placement. The sensor placement strategy is centralized at the cluster level since every cluster head makes redeployment decisions for the nodes in its cluster. Nevertheless, the clusters make deployment decisions independently, hence there is a considerable degree of decentralization in the overall sensor deployment. The virtual force in the VFA algorithm is calculated with a grid point being the location indicator and the distance between two grid points being a measure of distance. Furthermore, in our simulations, the preferential areas and the obstacles are both modeled as rectangles. The VFA algorithm however is also applicable for alternative location indicators, distance measures, and models of preferential areas and obstacles. Hence the VFA algorithm can be easily extended to heterogeneous sensors, where sensors may differ from each other in their detection modalities and parameters.

In Section 2.4, we have formulated an optimization problem on uncertainty-aware sensor placement. A minimum number of sensors are deployed to pro-

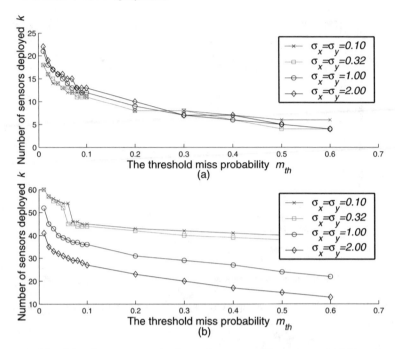

Fig. 2.39. Number of sensors required as a function of the miss probability threshold in presence of obstacles with $\alpha = 0.5, e^x_{max} = e^y_{max} = 5$ for (a) MIN_MISS (b) MAX_MISS.

vide sufficient grid coverage of the sensor field though the exact sensor locations are not known. The sensor location has been modeled as a random variable with a Gaussian probability distribution. We have presented two polynomial-time algorithms to optimize the number of sensors and determine their placement in an uncertainty-aware manner. The proposed algorithms address coverage optimization under constraints of imprecise detections and terrain properties.

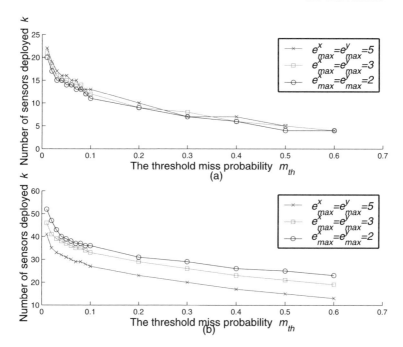

Fig. 2.40. Comparisons in different truncated Gaussian models in presence of obstacles with $\alpha = 0.5, \sigma_x = \sigma_y = 2$ for (a) MIN_MISS (b) MAX_MISS.

3

Energy-Aware Target Localization

Energy is a critical resource in wireless sensor networks and system lifetime needs to be prolonged through the use of energy-conscious sensing strategies during system operation. Energy management in these networks is crucial since battery-driven sensor nodes are severely energy-constrained. We focus here on reducing energy consumption in cluster-based wireless sensor networks for target localization.

In this chapter[1], we describe a target localization approach based on a two-step communication protocol between the cluster head and the sensors within the cluster[136, 137]. Since the energy consumption in wireless sensor networks increases significantly during periods of activity, which may be triggered for example by a moving target [12, 13], we propose an energy-reduction method for target localization in cluster-based wireless sensor networks. In the first step, sensors detecting a target report the event to the cluster head. The amount of information transmitted to the cluster head is limited; in order to save power and bandwidth, the sensor only reports the presence of a target, and it does not transmit detailed information such as signal strength, confidence level in the detection, imagery or time series data. Based on the information received from the sensor and the knowledge of the sensor deployment within the cluster, the cluster head executes a probabilistic scoring-based localization algorithm to determine likely position of the target. The cluster head subsequently queries a subset of sensors that are in the vicinity of these likely target positions.

In our two-step communication protocol, when a sensor detects a target, it sends an event notification to the cluster head. In order to conserve power and bandwidth, the message from the sensor to the cluster head is kept very small; in fact, the presence or absence of a target can be encoded in just one bit. Detailed information such as detection strength level, imagery and time series

---

[1] This chapter is based on Y. Zou and K. Chakrabarty, "Target localization based on energy considerations in distributed sensor networks", *Ad Hoc Networks*, vol. 1, pp. 261–272, 2003.

data are stored in the local memory and provided to the cluster head upon subsequent queries. Based on the information received from the sensors within the cluster, the cluster head executes a probabilistic localization algorithm to determine candidate target locations, and it then queries the sensor(s) in the vicinity of the target.

3.1 Detection Probability Table

The cluster head first generates a detection probability table for each grid point. The detection probability table contains entries for all possible detection reports from those sensors that can detect a target at this grid point. Let us assume the sensor field is represented by a $m \times n$ grid, and a grid point P at (i, j) is covered by a set of k_{ij} sensors, denoted as S_{ij}, $|S_{ij}| = k_{ij}, 0 \leq k_{ij} \leq k$, and $S_{ij} \subseteq \{s_1, s_2, \cdots, s_k\}$. The probability table is built on the power set of S_{ij} since there are $2^{k_{ij}}$ possibilities for k_{ij} sensors in reporting an event. These $2^{k_{ij}}$ cases include the event that none of the sensors detect anything (represented by the binary string as "00...0") as well as the event that all of the sensors (represented by the binary string as "11...1"). Thus the probability table for grid point (i, j) then contains $2^{k_{ij}}$ entries, defined as:

$$p\_table_{ij}(l) = \prod_{s_p \in S_{ij}} p_{ij}(s_p, l) \tag{3.1}$$

where $0 \leq l \leq 2^{k_{ij}}$, and $p_{ij}(s_p, l) = c_{ij}(s_p)$ if s_j detects a target at grid point $P(i, j)$; otherwise $p_{ij}(s_p, l) = 1 - c_{ij}(s_p)$. Table 3.1 gives an example of the probability tables on a 5 by 5 grid with 3 sensors deployed.

Table 3.1. Example probability table.

l	$d_1 d_2 d_3$	$p\_table_{ij}(l), \ 0 \leq l < 2^{k_{ij}}, \ k_{ij} = 3$
0	000	$(1 - 0.5736) \times (1 - 1) \times (1 - 0.5736) = 0.0$
1	001	$(1 - 0.5736) \times (1 - 1) \times 0.5736 = 0.0$
2	010	$(1 - 0.5736) \times 1 \times (1 - 0.5736) = 0.1819$
3	011	$(1 - 0.5736) \times 1 \times 0.5736 = 0.2446$
4	100	$0.5736 \times (1 - 1) \times (1 - 0.5736) = 0.0$
5	101	$(1 - 0.5736) \times (1 - 1) \times 0.5736 = 0.0$
6	110	$0.5736 \times 1 \times (1 - 0.5736) = 0.2446$
7	111	$0.5736 \times 1 \times 0.5736 = 0.3290$

Consider the grid point $(2, 4)$ in Figure 3.1 which is covered by all three sensors s_1, s_2 and s_3 with probabilities as $0.57, 1$, and 0.57 respectively. For the three sensors s_1, s_2 and s_3, there are a total of 8 possibilities for their

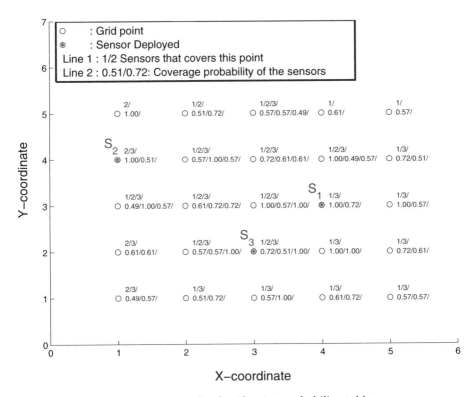

Fig. 3.1. Example of grid point probability table.

combined event detection at grid point $(2,4)$. For example, the binary string 110 denotes the possibility that s_1 and s_2 report a target but s_3 does not report a target. For each such possibility $d_1 d_2 d_3$ $(d_1, d_2, d_3 \in \{0, 1\})$ for a grid point, we calculate the conditional probabilities that the cluster head receives $d_1 d_2 d_3$ given that a target is present at that grid point. For our example, these conditional probabilities are listed in Table 3.1. Consider the binary string 110, the conditional probability associated with this possibility is given by $p\_table_{24}(6) = p_{24}(s_1, 6)p_{24}(s_2, 6)p_{24}(s_3, 6) = 0.57 \times 1 \times (1 - 0.57) = 0.24$. Note that the probability table generation is only a one-time cost. Once the probability table is generated, there is no need to refresh it unless sensor locations are changed.

3.2 Score-Based Ranking

After the probability table is generated for all the grid points, localization is done by the cluster head if a target is detected by one or more sensors. We use an inference method based on the established probability table. When at time instant t, the cluster head receives positive event message from $k(t)$ sensors, it uses the grid point probability table to determine which of these sensors are most suitable to be queried for more detailed information. Detailed target reporting consumes more energy consumption and it needs more bandwidth. Therefore, the cluster head cannot afford to query all the sensors for detailed reports. There is also an inherent redundancy in sensor detection information so it is not necessary to query all sensors. Our scoring approach is able to select the most suitable sensors for this purpose.

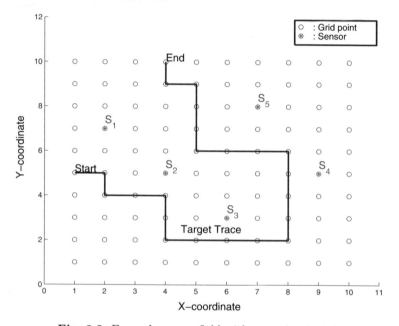

Fig. 3.2. Example sensor field with a moving target.

Consider the 10 by 10 grid shown in Figure 3.2. There are five sensors deployed, $k = 5$, $r = 2$ and $r_e = 1$. The zigzag shaped line is the target movement trace. The target starts to move at $t = t_{start}$ from the grid point marked as "Start" and finishes at $t = t_{end}$ at the grid point marked as "End". Figure 3.3 gives the score report at the time instant t_{start} when the target is present at "Start".

Assume $S_{rep}(t)$ is the set of sensors that have reported the detection of an object at time t, $S_{rep,ij}(t)$ is the set of sensors that can detect a target at point $P(i, j)$ and have also reported the detection of an object at time t. Obviously, $S_{rep,ij}(t) \subseteq S_{rep}(t)$ and $S_{rep,ij}(t) \subseteq S_{ij}$ since $S_{rep,ij}(t) = S_{rep}(t) \cap S_{ij}$. The score of the grid point $P(i, j)$ at time instant t is calculated as follows:

$$SCORE_{ij}(t) = p\_table_{ij}(l(t)) \times w_{ij}(t) \tag{3.2}$$

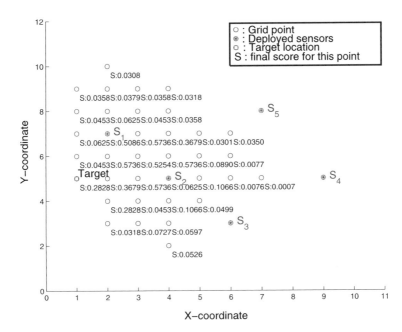

Fig. 3.3. Scoring results for target in the example sensor field at t_{start}. s_1 and s_2 have reported.

where $l(t)$ is the index of the p_table_{ij} at time t. The parameter $l(t)$ is calculated from S_{ij} and $S_{rep,ij}$. The parameter $p\_table_{ij}(l(t))$ corresponds to the conditional probability that the cluster head receives this event information given that there was a target at $P(i, j)$. The weight $w_{ij}(t)$ reflects the confidence level in this reporting event for this particular grid point. In our previous work [134], we have used the weight factor $w_{ij}(t) = \frac{k_{rep,ij}(t)}{k_{rep}(t)}$; this is sufficient for selecting sensors in order to conserve energy. However, in order to refine the grid point scores to narrow down grid points that are most probably close to the current target location, we have redefined $w_{ij}(t)$ here to improve the accuracy for target location. The weight for the grid point $P(i, j)$ at time instant t is defined as,

$$w_{ij}(t) = \begin{cases} 0 & \text{if } S_{rep,ij}(t) = \{\phi\} \\ 4^{-\Delta k_{rep,ij}(t)} & \text{otherwise} \end{cases} \tag{3.3}$$

where $\Delta k_{rep,ij}(t)$ measures the degree of difference in the set of sensors that reported and those sensors that can detect point $P(i, j)$ at time instant t. The parameter $\Delta k_{rep,ij}(t)$ is defined as

$$\Delta k_{rep,ij}(t) = |k_{rep}(t) - k_{rep,ij}(t)| + |k_{rep}(t) - k_{ij}| \tag{3.4}$$

where $k_{ij} = |S_{ij}|$, $k_{rep}(t) = |S_{rep}(t)|$, and $k_{rep,ij}(t) = |S_{rep,ij}(t)|$. The parameter $w_{ij}(t)$ is therefore a decaying factor that is 1 only if $S_{rep}(t) = S_{ij}$. The number

4 in the formula for $w_{ij}(t)$ was chosen empirically after it was found to provide accurate simulation results. We are using $w_{ij}(t)$ to filter out grid points that are not likely to be close to the actual target location. The score is based on both the probability value from the probability table and the current relationship between $S_{rep}(t)$, $S_{rep,ij}(t)$ and S_{ij}. Table 3.2 gives some score calculation examples for the grid points in Figure 3.3 at the time instant t_{start}.

Table 3.2. Scoring calculation example for t at t_{start}.

(x,y)	S_{ij}	$S_{rep,ij}(t)$	$w_{ij}(t)$	$p\_table_{ij}(l(t))$	$SCORE_{ij}(t)$
$\cdots$	$\cdots$	$\cdots$	$\cdots$	$\cdots$	$\cdots$
$(1,6)$	s_1	s_1	0.25	0.7248	0.0453
$(2,6)$	s_1, s_2	s_1, s_2	1.00	0.5736	0.5736
$(3,6)$	s_1, s_2	s_1, s_2	1.00	0.5254	0.5254
$(4,6)$	s_1, s_2	s_1, s_2	1.00	0.5736	0.5736
$(5,6)$	s_2, s_5	s_2	0.25	0.3562	0.0890
$(6,6)$	s_2, s_3, s_5	s_2	0.25	0.1240	0.0077
$\cdots$	$\cdots$	$\cdots$	$\cdots$	$\cdots$	$\cdots$

3.3 Selection of Sensors to Query

Assume that the maximum number of sensors that are allowed to report an event is k_{max}, and the set of the sensors selected by the cluster head for querying at time t is $S_q(t), S_q(t) \subseteq S_{rep}(t) \subseteq \{s_1, s_2, \cdots, s_k\}$. To select the sensor to query based on the event reports and the localization procedure, we first note that for time instant t, if $k_{max} \geq k_{rep}(t)$, then all reported sensors can be queried. Otherwise, we select sensors based on a score-based ranking. The sensors selected correspond to the ones that have the shortest distance to those grid points with the highest scores. This selection rule is defined as:

$$S_q(t) : d(S_q(t), P_{MS}) = min\{d(s_i, P_{MS})\} \tag{3.5}$$

where $s_i \in S_{rep}(t)$, and P_{MS} denotes the set of grid points with the highest scores. Note it is possible that there are multiple grid points that have the maximum score. When this happens, we calculate the score concentration by averaging the scores of the current grid point and its eight neighboring grid points. The grid point with the highest score (or the score concentration) is the most likely current target location. Therefore, selecting sensors that are closest to this point guarantee that the selected sensors can provide the most detailed and accurate data in response to the subsequent queries. Note target identification is not possible as at this stage since the cluster head has no additional information other than $S_{rep}(t)$. However, the selected sensors provide enough information in the subsequent stage to facilitate target identification. We evaluate the accuracy of this target localization procedure by calculating the distance between the grid point with the highest score and the actual target location. For the example of Figure 3.2, Table 3.3 gives some results for the selected sensor when the target is moving from "Start" ($t = 1$) to "End". We assume $k_{max} = 1$, and the target is moving at a constant speed. $\bar{S}_q(t)$ is the set of sensors that are closest to the actual location of the target at time t. The results show that $S_q(t)$ matches $\bar{S}_q(t)$ in many cases. The example does not to illustrate the advantages of our proposed strategy since not many sensors are actually involved at the same time for target detection. However, we show later in Section 3.6 that the proposed algorithm performs very well when many sensors are involved in the target detection and reporting process.

Table 3.3. Selected sensors for the example in Figure 3.2.

t	$S_{rep}(t)$	$S_q(t)$	$\bar{S}_q(t)$	t	$S_{rep}(t)$	$S_q(t)$	$\bar{S}_q(t)$
...	...	...	...	...	...	...	...
3	s_1, s_2	s_1	s_2	4	s_2	s_2	s_2
5	s_2, s_3	s_3	s_2	6	s_2, s_3	s_3	s_2
7	s_2, s_3	s_3	s_3	8	s_3	s_3	s_3
...	...	...	...	...	...	...	...
16	s_4, s_5	s_4	s_4	17	s_4, s_5	s_4	s_5
18	s_2, s_3, s_5	s_2	s_2	19	s_2, s_5	s_5	s_2
20	s_1, s_2, s_5	s_2	s_2	21	s_5	s_5	s_5
...	...	...	...	...	...	...	...

3.4 Energy Evaluation Model

Let us consider the energy consumption for a sensor network that is actively detecting a target in the sensor field. We assume sensor nodes are homogeneous, therefore the energy consumption for sensing is the same for each sensor node. To focus on energy consumption due to target activities or events, this work does not consider the energy consumed by sensor nodes when they are in the idle state. This does not however imply that the energy consumption of idle sensor nodes can always be ignored.

3.4.1 Primitive Energy Evaluation Model

To simplify the energy analysis, let us first consider a primitive sensor model, which focuses on the energy consumption of the wireless sensor network due to the target activities or events. Suppose the sensor node has three basic energy consumption types—sensing, transmitting and receiving, and these power values (energy per unit time) are E_s, E_t and E_r, respectively. If we select all sensors that reported the target for querying, the total energy consumed for the event happening at time instant t can be evaluated using the following set of equations:

$$E_1(t) = k_{rep}(t)(E_t + E_r)T_1 \tag{3.6}$$

$$E_2(t) = (k_{rep}(t)E_r + E_t)T_2 \tag{3.7}$$

$$E_3(t) = k_{rep}(t)(E_t + E_r)T_3 \tag{3.8}$$

$$E_4(t) = E_s T_s \tag{3.9}$$

$$E(t) = E_1(t) + E_2(t) + E_3(t) + E_4(t) \tag{3.10}$$

$$E = \sum_{t=t_{start}}^{t_{end}} E(t) \tag{3.11}$$

where E_1 is the energy required for reporting the detection of an object, E_2 is the energy required for transmitting query information from the cluster head by broadcasting and for receiving this information at the sensor nodes, and E_3 is the energy required by sensor nodes being queried to send detailed information to the cluster head. The parameters T_1, T_2 and T_3 denote the lengths of time involved in the transmission and reception, which are directly proportional to the sizes of data for yes/no messages, control messages to query sensors, and the detailed sensor data transmitted to the cluster head. The parameter T_s is the time of sensing activity of sensors. The parameters E denotes the total energy in this case for target localization from t_{start} to t_{end}. For the proposed probabilistic localization approach, we calculate the total energy consumption $E^\star$ as follows:

$$E_1^*(t) = k_{rep}(t)(E_t + E_r)T_1 \tag{3.12}$$

$$E_2^*(t) = (k_q(t)E_r + E_t)T_2 \tag{3.13}$$

$$E_3^*(t) = k_q(t)(E_t + E_r)T_3 \tag{3.14}$$

$$E_4^*(t) = E_s T_s \tag{3.15}$$

$$E^*(t) = E_1^*(t) + E_2^*(t) + E_3^*(t) + E_4^*(t) \tag{3.16}$$

$$E^* = \sum_{t=t_{start}}^{t_{end}} E^*(t) \tag{3.17}$$

where $E_1(t)^\star = E_1(t)$, $E_4^\star(t) = E_4(t)$, and the total energy consumed is denoted by E^*. Therefore, the energy savings via the use of the probabilistic target localization algorithm is:

$$\Delta E = E - E^* = C \sum_{t=t_{start}}^{t_{end}} (k_{rep}(t) - k_q(t)) \tag{3.18}$$

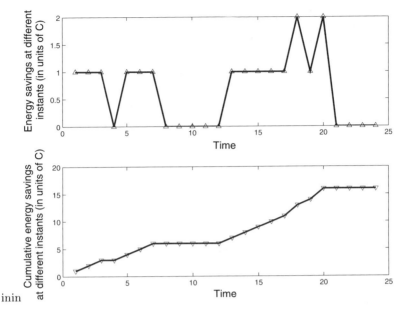

Fig. 3.4. Energy saved for the example in Figure 3.2 using the primitive energy evaluation model.

where $C = E_r T_2 + (E_t + E_r) T_3$ is a constant. Since $k_q(t)$ is always less than or equal to $k_{rep}(t)$, we have $\Delta E \geq 0$. Also, ΔE is monotonically non-decreasing with time. Figure 3.4 shows the energy saved for the target trace in Figure 3.2.

3.4.2 Refined Energy Evaluation Model

The previous primitive energy evaluation model given by Equation (3.6) to (3.18) convey the idea that large volume of data can greatly burden the energy consumption on sensor nodes. We therefore refine the energy evaluation model to incorporate the consideration of the overhead induced by our approach. This refined model is used later as the major metric for evaluating energy consumption with sample testing values from [12, 52, 104]. We still consider a sensor node with three basic energy consumption types—sensing, transmitting and receiving, and these power values (joule per second) are ψ_s, ψ_t and ψ_r, respectively. Assume at time instant t, there are $k(t)$ sensors that have detected the target, where $k(t) \leq k$. Therefore, the energy for sensing activities in the wireless network, denoted as $E_s(k(t))$, is

$$E_s(k(t)) = k(t)\psi_s T_s \tag{3.19}$$

where T_s is the time duration that a sensor node is involved in sensing. For a fixed time interval, E_s is a constant if all sensor nodes are assumed to be homogenous. The energy used for communication between nodes and the cluster head can be categorized into two types, E_b and E_c. The parameter E_c is the energy consumed by a sensor node for communication with the cluster head. This includes the energy for transmitting data and the energy for receiving data. The parameter E_b is the energy needed for broadcasting data from the head to the nodes. Both E_b and E_c are functions of T and $k(t)$, where T is the time required for either retrieving data from a sensor node or the broadcasting of data from the cluster head, and $k(t)$ is the number of sensors involved in this communication at time instant t. We define E_c and E_b as follows:

$$E_c(k(t), T) = (\psi_t T + \psi_r T)k(t) \tag{3.20}$$
$$E_b(k(t), T) = \psi_t T + \psi_r T k(t) \tag{3.21}$$

The parameter T is directly proportional to the volume of data involved in the communication. In this work, T can be one of three values, T_d for raw target data, T_e for target event reporting, and T_q for query request. They satisfy the relationship $T_e \leq T_q \ll T_d$ since raw data collected by a sensor node can be up to hundreds of bytes in size. We assume that target detection and localization are discrete processes, which are derived from a discrete sampling of target activities in the sensor network. Also, since the sensor network is designed to track target activities, T_s, T_e, T_q, and T_d are assumed to be less than the granularity of the time t. Thus for the case that a target is moving in the sensor field during the time interval $[t_{start}, t_{end}]$, the corresponding instantaneous energy consumption $E(t)$ and total energy consumption E in the wireless sensor network can be expressed as

$$E(t) = E_s(k(t)) + E_c(k(t), T_d) \tag{3.22}$$

$$E = \sum_{t=t_{start}}^{t_{end}} E(t). \tag{3.23}$$

From Equation (3.19) to Equation (3.23), we evaluate the energy consumption using the above target localization method as follows:

$$E^*(t) = E_s(k_{rep}(t)) + E_c(k_{rep}(t), T_e) + E_b(k_q(t), T_q) + \tag{3.24}$$
$$E_c(k_q(t), T_d) \tag{3.25}$$

$$E^* = \sum_{t=t_{start}}^{t_{end}} E^*(t) \tag{3.26}$$

Therefore, let $k(t) = k_{rep}(t)$ in Equations (3.22) and (3.23), the difference in energy consumption, $\Delta E = E - E^*$ can be expressed as;

$$\Delta E(t) = (k_{rep}(t) - k_q(t))(\psi_t + \psi_r)T_d$$
$$- (k_q(t)\psi_r + \psi_t)T_q - k_{rep}(t)(\psi_t + \psi_r)T_e \tag{3.27}$$

$$\Delta E = \sum_{t=t_{start}}^{t_{end}} \Delta E(t) \qquad (3.28)$$

The last two terms in Equation (3.27) indicate the overhead for the proposed target localization procedure. Since $T_d \gg T_e$, and $T_d \gg T_q$, the overhead is small. As $k_q < k_{max}$, with k_{max} properly selected, from Equation (3.27) and Equation (3.28), energy consumption is greatly reduced with the passage of time.

3.5 Procedural Description

Figure 3.5 shows the pseudocode of the procedure to generate the detection probability table for each grid point. Figure 3.6 shows the pseudocode for the simulation of the probabilistic localization algorithm. For an n by m grid with k sensors, the computational complexity involved in generating the probability table is $O(nm2^k)$ since the maximum number of sensors that can detect a grid point is k for the worst case. The computational complexity of the localization procedure is $O(nmk_{max})$, $k_{max} \leq k$. Therefore, the computational complexity of the probabilistic localization algorithm is $\max\{O(nmk_{max}), O(nm2^k)\} = O(nm2^k)$. Even though the worst-case com-

---

Procedure *Generate_Probability_Table* $(\text{P(i,j)}, \{\text{s}_1, \cdots, \text{s}_k\})$

---

1 /* find S_{ij}, the set of sensors that can detect $P(i,j)$ */
2 **For** $s_p \in \{s_1, s_2, \cdots, s_k\}$
3   **If** $d_{ij}(s_p) \leq r + r_e$
4      $S_{ij} = S_{ij} \cup \{s_p\}$;
5   **End**
6 **End**
7 /* fill up the probability table */
8 **For** $l, 0 \leq l \leq k_{ij}, k_{ij} = |S_{ij}|$;
9   **If** s_p detects $P(i,j)$
10   Set $p_{ij}(s_p, l) = c_{ij}(s_p)$;
11   **Else**
12   Set $p_{ij}(s_p, l) = 1 - c_{ij}(s_p)$;
13   **End**
14   Set $p\_table_{ij}(l) = \prod_{s_p \in S_{ij}} p_{ij}(s_p, l)$;

15 **End**

Fig. 3.5. Pseudocode for generating the detection probability table.

plexity of the localization procedure is exponential in k, in practice, the localization procedure can execute in less time since the number of sensors that can effectively detect a target at a given grid point is quite small.

Procedure *Target_Localization*(**Grid**, $\{s_1, ..., s_k\}$, **TargetTrace**)

/* k_{max} is the maximum number of sensors that are allowed for querying,
p_{rep} is the threshold level for a sensor to report to
the cluster head of an event. $TargetTrace$ starts from t_{start} and
ends at t_{end}, with time unit as 1. */
1 Set $t = t_{start}$;
2 **While** $(t \leq t_{end})$
3   /* current target location */
4   Set $Target = TargetTrace(t)$;
5   /* calculate the scores */
6   Calculate $S_{rep}(t)$ from $\{s_1, s_2, \cdots, s_k\}, Target(t), p_{rep}$;
7   Set $k_{rep}(t) = |S_{rep}(t)|$;
8   **For** $P(i, j)$ in Grid, $i \in [1, width], j \in [1, height]$
9   Set $k_{ij} = |S_{ij}|$;
10   Calculate $S_{rep,ij}(t)$ from $S_{rep}(t)$ and $P(i, j)$;
11   Calculate the index $l(t)$ of p_table_{ij} from
              $S_{rep}(t)$, and $S_{rep,ij}(t)$;
12   Set $k_{rep,ij}(t) = |S_{rep,ij}(t)|$;
13   **If** $S_{rep,ij}(t) = \{\phi\}$
14              $w_{ij}(t) = 0$;
15   **Else**
16   Set $\Delta k_{rep,ij}(t) = |k_{rep}(t) - k_{rep,ij}(t)|$
                  $+ |k_{rep}(t) - k_{ij}|$;
17              $w_{ij}(t) = 4^{-\Delta k_{rep,ij}(t)}$;
18   **End**
19   Set $SCORE_{ij}(t) = p\_table_{ij}(l(t)) \times w_{ij}(t)$;
20   **End**
21   /* select sensors for querying */
22   Calculate $S_q(t)$ from $SCORE_{ij}(t)$ and k_{max}, $i \in [1, width], j \in [1, height]$;
23   /* next time instant */
24   Set $t = t + 1$;
25 **End**

Fig. 3.6. Pseudocode of the target localization procedure.

3.6 Simulation Results

In this section, we present results for three case studies carried out on a Pentium III 1.0 GHz PC using Matlab.

3.6.1 Case Study 1

We first evaluate the localization algorithm using the results produced by the VFA algorithm in the sensor deployment stage introduced in Chapter 2. At this point, sensors are already moved to proper locations by the VFA algorithm. Figure 3.7 shows the sensor locations. There are total of 20 sensors deployed on a 50×50 sensor field grid, $r = 5$ grid units, $r_e = 3$ grid units, $c_{th} = 0.7$, $\lambda = 0.5$, and $\beta = 0.5$. To simulate target movement, we consider a target movement trace in the sensor grid as shown in Figure 3.7. The parameter t_{start} is the time instant that the target starts to move from its initial location marked as "Start" in Figure 3.7. Table 3.4 shows the results of the localization algorithm. We assume that a maximum of two sensors can be selected for querying by the cluster head. The target is assumed to move only 1 grid unit in one unit of time. There are total of 82 such moves in the simulated target movement trace. In the interest of conciseness, we only present the results for moves numbered 1-5, 41-45 and 78-82.

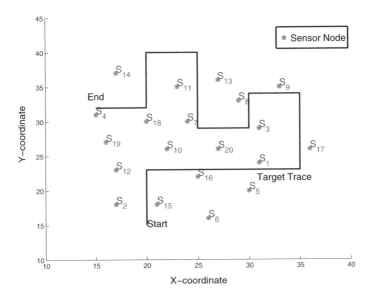

Fig. 3.7. Sensor field with sensors deployed by the VFA algorithm and a target movement trace.

The set $S_{rep}(t)$ indicates sensors that have reported the detection at time instant t. The set $S_q(t)$ includes sensors that are selected for querying by the cluster head at time t. The parameter $\Delta E(t)$ shows the energy saved by the localization algorithm for the detection event at time instant t. Figure 3.8 shows the estimated target location based on the grid point with the highest

Table 3.4. Sensors selected for querying by the cluster head.

t	$S_{rep}(t)$	$S_q(t)$	$\bar{S}_q(t)$	P_{MS}	$Target$	$\Delta E(t)$
01	s_2, s_6, s_{15}	s_2, s_{15}	s_2, s_{15}	$(21, 15)$	$(20, 15)$	C
02	$s_2, s_6, s_{12}, s_{15}, s_{16}$	s_2, s_{12}	s_2, s_{15}	$(21, 15)$	$(20, 15)$	$3C$
03	$s_2, s_6, s_{12}, s_{15}, s_{16}$	s_2, s_{12}	s_2, s_{15}	$(21, 15)$	$(20, 15)$	$3C$
04	$s_2, s_6, s_{12}, s_{15}, s_{16}$	s_2, s_{12}	s_2, s_{15}	$(21, 15)$	$(20, 15)$	$3C$
05	$s_2, s_6, s_{12}, s_{15}, s_{16}$	s_2, s_{12}	s_2, s_{15}	$(21, 15)$	$(20, 15)$	$3C$
...	...	...	...	...	...	...
41	s_3, s_8, s_9, s_{13}	s_8, s_9	s_8, s_9	$(31, 34)$	$(31, 34)$	$2C$
42	$s_3, s_7, s_8, s_9, s_{11}, s_{13}$	s_8, s_{13}	s_8, s_9	$(28, 34)$	$(30, 34)$	$4C$
43	$s_3, s_7, s_8, s_9, s_{11}, s_{13}, s_{20}$	s_{13}, s_8	s_8, s_9	$(27, 34)$	$(30, 33)$	$5C$
44	$s_3, s_7, s_8, s_9, s_{11}, s_{13}, s_{20}$	s_{13}, s_8	s_8, s_3	$(27, 34)$	$(30, 32)$	$5C$
45	$s_3, s_7, s_8, s_9, s_{11}, s_{13}, s_{20}$	s_{13}, s_8	s_8, s_3	$(27, 34)$	$(30, 32)$	$5C$
...	...	...	...	...	...	...
78	$s_4, s_7, s_{10}, s_{11}, s_{14}, s_{18}, s_{19}$	s_{18}, s_7	s_{18}, s_4	$(27, 34)$	$(30, 32)$	$5C$
79	$s_4, s_7, s_{10}, s_{11}, s_{14}, s_{18}, s_{19}$	s_{18}, s_7	s_{18}, s_4	$(27, 34)$	$(30, 32)$	$5C$
80	$s_4, s_7, s_{10}, s_{11}, s_{14}, s_{18}, s_{19}$	s_{18}, s_7	s_4, s_{18}	$(27, 34)$	$(30, 32)$	$5C$
81	$s_4, s_{11}, s_{14}, s_{18}, s_{19}$	s_4, s_{18}	s_4, s_{18}	$(27, 34)$	$(30, 32)$	$3C$
82	$s_4, s_{14}, s_{18}, s_{19}$	s_4, s_{19}	s_4, s_{19}	$(27, 34)$	$(30, 32)$	$2C$

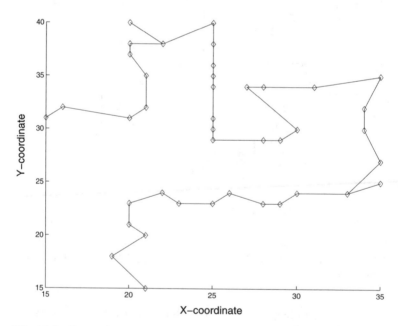

Fig. 3.8. Target localization by the grid point with the highest score.

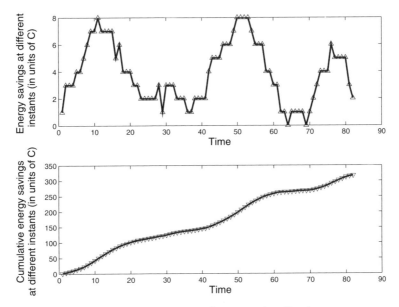

Fig. 3.9. Energy saving for target localization.

score. Figure 3.9 shows the energy saved during the target tracking process. Energy saved is evaluated in units of the constant C, given by Equation (3.18) in Section 3.4.1. The total computation time for generating the probability table is only 11 seconds. The total computation time for target localization for the total of 82 locations is only 16 seconds, with an average of 0.2 second per time instant.

3.6.2 Case Study 2

We next present the simulation on a 30 by 30 sensor field grid with 20 sensors randomly placed in the sensor filed. The parameters of the sensor detection model are $r = 5$, $r_e = 4$, $\lambda = 0.5$, and $\beta = 0.5$. We choose energy the consumption model parameters as $\psi_r \approx 400$ nJ/sec, $\psi_t \approx 400$ nJ/sec, and $\psi_s \approx 1000$ nJ/sec. These values are based on the typical values given in [12, 52, 104], assuming the sensing rate for a the sensor is 8 bits/sec. We have no physical data available for T_d and T_e; however, their values do not affect the target localization procedure, therefore we only need to set them manually to satisfy the relationship $T_d \gg T_e$ and $T_d \gg T_q$. In this case, $T_d = 100$ ms, $T_e = 2$ ms, and $T_q = 4$ ms.

The layout of the sensor field is given in Figure 3.10, with a target trace randomly generated in the sensor field. The target travels from the position marked as "Start" to the position marked as "End". We assume the target locations are updated at discrete time instants in unit of seconds, and the granularity of time is long enough for sampling by two neighboring locations

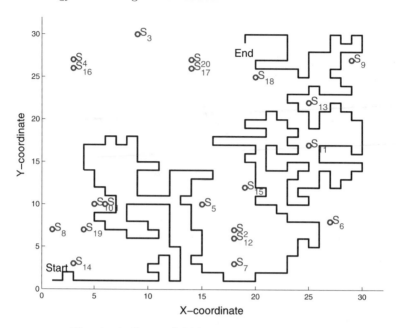

Fig. 3.10. Sensor field layout with target trace.

in the target trace with negligible errors. We have evaluated the algorithm for $k_{max} = 1$, $k_{max} = 2$, and $k_{max} = 3$.

Figure 3.11 presents the instantaneous energy saving in percentage, and Figure 3.12 presents the absolute value of the cumulative energy saving for the case study as the target moves along its trace in the sensor field. The energy savings are compiled relative to the base case when all sensors report complete target information in one step everywhere.

From Figure 3.11 and Figure 3.12, we note that a large amount of energy is saved during target localization. Note that when k_{max} approaches $k_{rep}(t)$, the saving is less apparent due to the additional communication overhead of the two-stage query protocol. Nevertheless, there is still a considerable amount of energy saved in target localization, even for the case that $k_{max} = 3$. With an appropriate selection of k_{max}, the proposed algorithm performs exceptionally well.

Next, we consider the latency in the localization of a target by the cluster head. By latency, we refer to the time that it takes for the cluster head to collect the detailed target information from sensor nodes from the time sensor nodes detect an event, assuming that the wireless sensor network uses the time division multiple access (TDMA) protocol [122]. The results are shown in Figure 3.13. The latency is reduced here compared to the base case using a "report once" strategy, since a large amount of communication for transmitting raw data has been reduced to a smaller amount of data sent by a selected set

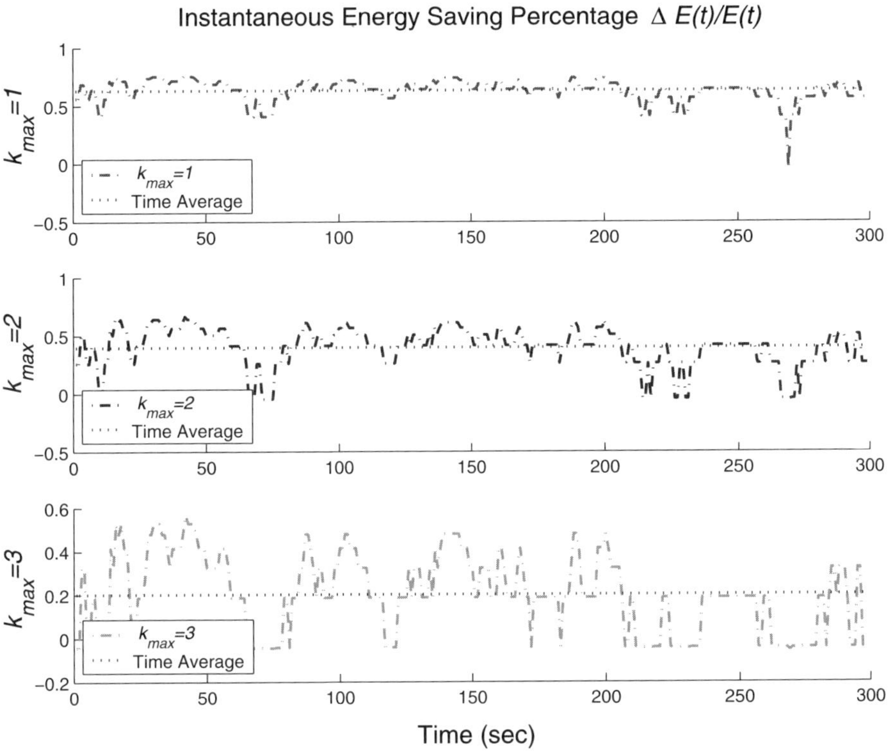

Instantaneous Energy Saving Percentage $\Delta\,E(t)/E(t)$

Fig. 3.11. Instantaneous energy saving percentage during target localization relative to the "always report" one-step base case.

of sensors. This is an added advantage to the proposed energy-aware target localization procedure.

Since the selection of sensors for querying is based on both the detection probability table and the distance of sensors from the estimated high-score points, the proposed *a posteriori* approach offers another important advantage. It provides a substantial amount of built-in false-alarm filtering. Figure 3.14 illustrate the false-alarm filtering ability of the proposed approach. We manually generated false alarms reported by some malfunctioning sensors, which are during $t \in [18, 22]$ by s_4, during $t \in [138, 142]$ by s_{16}, and during $t \in [239, 241]$ by s_8. We calculate the distance d of the target from the sensor in $S_{rep}(t)$ that is farthest from it, as well as the distance d^* of the target from the sensor in $S_q(t)$ that is farthest from it. The difference $d - d^*$ is used as a measure of the built-in filtering ability. Figure 3.14 shows the variation of $d - d^*$ with time. Note the fact that prior to querying, the cluster head only knows which sensors have reported the detection of a target, and there is no information available to the cluster head about any detailed information of the target.

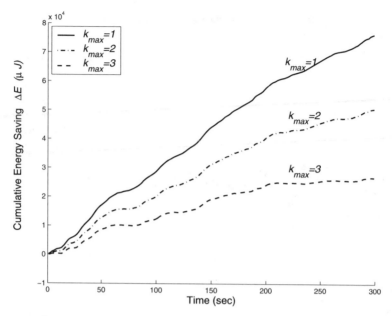

Fig. 3.12. Cumulative energy saved during target localization relative to the "always report" one-step base case.

We find that the proposed approach successfully narrows down the sensors that are the close to the real target location, and selects them for detailed information querying. As shown in Figure 3.14, the three spikes present the fact that the false alarms from the sensor (which in this case is the furthest sensor from the actual target location) have been filtered out since the proposed target localization procedure is still able to select the most appropriated sensors to query for detailed target information.

3.6.3 Case Study 3

Next, we present simulation results for the case of a target appearing at random locations in the sensor field with a constant event rate. The layout of the sensor field is the same as given by Figure 3.10. We generate a sequence of 200 random target locations on the sensor field to simulate a target appearing randomly for 200 discrete time instants. Simulation is carried out for $k_{max} = 1$, $k_{max} = 2$, and $k_{max} = 3$, with different sets of randomly generated target activity events at 200 discrete time instants. Figure 3.15 presents the instantaneous energy saving in percentage, and Figure 3.16 presents the absolute value of the cumulative energy saving.

Figure 3.17 presents the latency evaluation results for target localization. The proposed target localization strategy performs extremely well since en-

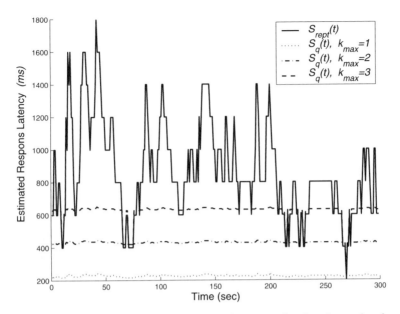

Fig. 3.13. Latency in the localization of a target by the cluster head.

ergy is saved for the case of random target appearance events, and latency is reduced for target localization.

3.7 Discussion

We have described an energy-aware target localization procedure for cluster-based wireless sensor networks. The proposed approach is based on the combination of a two-step communication protocol between the cluster head and the sensors in the cluster, and a probabilistic localization algorithm. We have shown that this approach reduces energy consumption, decreases the latency for target localization, and provides a mechanism for filtering false alarms. Future uture work needs to focus on the scalability of these algorithms. In particular, an interesting direction is to extend this approach for the localization of multiple targets as well as determining an appropriate value of the parameter k_{max}. A larger value of k_{max} is desirable for more accurate target classification, but the need for additional information for classification must be balanced with the need for low energy, reduced latency, and bandwidth requirements. Researchers can also consider more practical scenarios, e.g., when a sensor is temporarily blocked by external obstacles. These scenarios must be appropriately modeled in the sensor detection probability table.

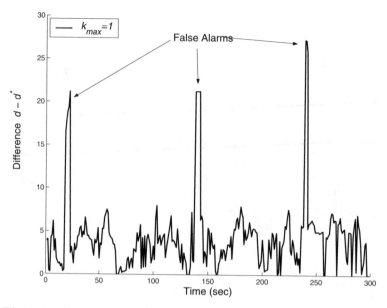

Fig. 3.14. Results on localization error in the presence of false alarms.

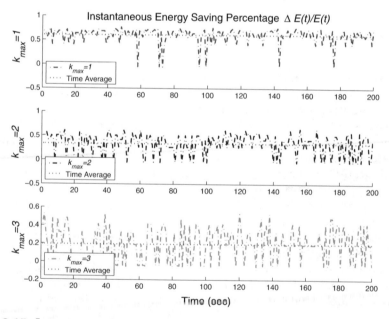

Fig. 3.15. Instantaneous energy saving percentage during target localization relative to the "always report" one-step base case.

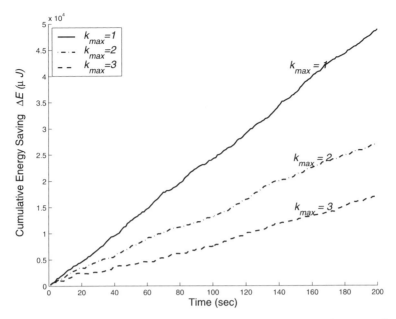

Fig. 3.16. Cumulative energy saved during target localization relative to the "always report" one-step base case.

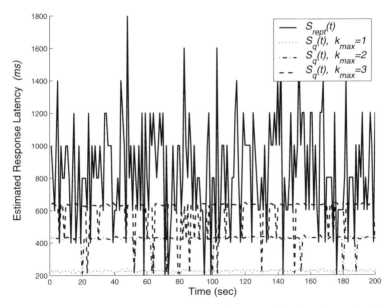

Fig. 3.17. Latency in the localization of a target by the cluster head.

4

Energy-Efficient Self-Organization

4.1 Introduction

Advances in miniaturization of microelectronic and mechanical structures (MEMS) have led to battery-powered sensor nodes that have sensing, communication and processing capabilities. Wireless sensor networks are networks of large numbers of such sensor nodes. Example applications of such sensor networks include the monitoring of wildfires, inventory tracking, assembly line monitoring, and target tracking in military systems.

Upon deployment in a remote or a hostile location, sensor nodes might fail with time due to loss of battery power, an enemy attack, or a change in environmental conditions. The replacement of each failed sensor node with a new sensor node is expensive and often infeasible, and it is therefore undesirable. Hence in such cases, a large number of redundant sensor nodes are deployed with the expectation that these nodes will be used later when some other nodes fail. The self-configuration of a large number of sensor nodes requires a distributed solution. In this chapter, we present a scalable self-configuration and an adaptive reconfiguration (SCARE) algorithm for distributed sensor networks.

An effective self-configuration scheme should have the following characteristics. It should be completely distributed and localized because a centralized solution is often not scalable for wireless sensor networks. It should be simple without excessive message overhead because sensor nodes typically have limited energy resources. It should be energy-efficient and require only a small number of nodes to stay awake and perform multi-hop routing, and it should keep the other nodes in a sleep state.

We describe a solution that meets the above design requirements [107]. We present a distributed self-configuration scheme that distributes the set of nodes in the sensor network into subsets of coordinator nodes and non-coordinator nodes. While coordinator nodes stay awake, provide coverage, and perform multi-hop routing in the network, non-coordinator nodes go to sleep. When nodes fail, SCARE adaptively re-configures the network by selecting appropriate non-coordinator nodes to become coordinators and take over the role of failed coordinators. This scheme only needs local topology information and uses simple data structures in its implementation.

The remainder of the chapter is organized as follows. In the next section we discuss related prior work. In Section 3, we present an overview of self-configuration using SCARE. Section 4 presents the details about SCARE, including the timeout and defer rules, and Section 5 presents an optimal centralized algorithm. The performance evaluation of SCARE is presented in Section 6. This section describes the simulation methodology, the experimental results and a comparison with related work. Finally, conclusions are presented in Section 7.

4.2 Relevant Prior Work

A number of topology management algorithms have been proposed for ad-hoc and sensor networks [4, 26, 109, 133]. While the connectivity problem has been studied in considerable detail for wireless ad hoc networks, less attention has been devoted to the problem of balancing connectivity and coverage. The GAF scheme [133] uses geographic location information of the sensor nodes and it divides the network into fixed-size virtual square grids. GAF identifies redundant nodes within each virtual grid and switches off their radios to achieve energy savings. In contrast, SCARE achieves energy savings by selectively powering down some of the nodes that are within the sensing radius of a coordinator.

A coverage-preserving node scheduling scheme is described in [119] that extends the LEACH [52] protocol to achieve energy savings. In this scheme, nodes advertise their position information in each round. Each node evaluates its eligibility to switch itself off by calculating its sensing area and comparing it with its neighbors's. If a node's sensing area is embraced by a union set of its neighbors's, then it turns itself off. To prevent blindspots in coverage due to several eligible nodes switching themselves off simultaneously, a back-off based scheduling is used. After the back-off interval has elapsed, nodes broadcast a status advertisement message to let other nodes know about their on/off status. Thus, each node broadcasts two messages in this scheme. In contrast, SCARE needs fewer than two messages per node on average during its operation. The scheme in [119] also utilizes location information of the nodes for its operation. SCARE only needs an estimate of the distance between the nodes.

The STEM scheme described in [109] trades off latency for energy savings by putting nodes aggressively to sleep and waking them up only when there is data to forward. It uses a second radio operating at a lower duty cycle for transmitting periodic beacons to wakeup nodes when there is data to forward. SCARE does not use a separate paging channel for self-configuration. Nevertheless, SCARE can integrate well with STEM to achieve significant energy savings.

In AFECA [132], nodes listen to the channel for transmissions. AFECA conservatively tries to keep nodes awake when there are not too many neighbors in its radio range. In order to deduce this information, each node has to listen to transmissions that are not meant for it. In SCARE, however, nodes listen at only periodic intervals in order to determine their states.

The PAMAS [101] multi-access protocol saves power by switching off the radio of a node when it is not transmitting or receiving. This method saves power when idle listening consumes significantly less energy compared to message reception.

The Span approach [26] appears to be the most closely related to SCARE. Span attempts to save energy by switching off redundant nodes without losing the connectivity of the network. Nodes make decisions based on their local topology information. However, SCARE differs from Span in that it uses distance estimates to determine the state of a node. Span uses a communication mechanism to obtain this information. Since Span was developed for ad hoc networks, its main focus is on ensuring network connectivity through energy-efficient topology management. It is not directed towards ensuring the sensing coverage of a given region. SCARE also differs from Span in that, in addition to ensuring network connectivity and low-energy self-configuration, it attempts to provide a high level of sensing coverage.

A TDMA-based self organization scheme for sensor networks is presented in [114]. Each node uses a superframe, similar to a TDMA frame, to schedule different time slots for different neighbors. However, this scheme does not take advantage of the redundancy inherent in wireless sensor networks to power off some nodes.

SCARE utilizes a localization scheme for periodic transmission of beacon signals and for the synchronization of the clock signals of sensor nodes. A number of such localization schemes have been proposed in the literature for sensor networks [9, 20, 44]. These schemes use a special set of nodes, called the reference nodes, that transmit beacon signals to let the sensor nodes self-estimate their position. The approach in [9] is based on the received signal strength from the reference nodes to carry out location estimation of the sensor nodes. It is shown that despite fading and mobility, a small window average is sufficient to do location estimation.

Traditionally, global positioning system (GPS) [56] receivers are used to estimate positions of the nodes in mobile ad-hoc networks. However, their high cost and the need for more precise location estimates make them unsuitable for sensor networks. It is expensive to add GPS capability to each device in dense sensor networks.

In [90], a scheme is presented to estimate the relative location of nodes using only a few GPS-enabled nodes. It uses the received signal strength information (RSSI) as the ranging method. An ad-hoc localization technique called *Calamari* is used in [125] in combination with a calibration scheme to calculate distance between two nodes using a fusion of RF-based RSSI and acoustic time of flight (TOF). Acoustic ranging [44] can also be used to get fine-grained position estimates of nodes.

Finally, several clustering techniques have been proposed in the ad hoc networking literature. A scheme is proposed in [121] that attempts to find maximal cliques in the physical topology, and uses a three-pass algorithm to find the clusters. Although this scheme finds a connected set of clusters, it consumes a significant amount of energy during clustering and cannot be directly applied to sensor networks. The adaptive clustering scheme proposed in [43] uses node IDs to build two-hop clusters in a deterministic manner. SCARE differs from this scheme in two ways. First, the main goal of SCARE is to use distance information to power down redundant sensor nodes, whereas in [43], node IDs are used to provide better QoS guarantees by clustering nodes. Second, in [43], as in [121], energy efficiency is a secondary concern. In [7], clustering schemes for both static and mobile networks are proposed. However, there is no provisioning for switching off redundant nodes in these schemes. Thus, [7] cannot be directly applied to sensor networks. On the other hand, SCARE is specifically designed for sensor networks to take advantage of their inherent redundancy.

4.3 Outline of SCARE

SCARE is a decentralized algorithm that distributes all the nodes in the network into subsets of coordinator nodes and non-coordinator nodes. While the coordinator nodes stay awake and provide coverage and perform multi-hop routing in the network, non-coordinator nodes go to sleep. Non-coordinator nodes wake-up periodically to check if they should become coordinators to replace failed coordinators.

SCARE achieves four desirable goals. First, it saves energy by selecting only a small number of nodes as coordinators and putting other nodes to sleep. Second, it uses only local topology information for coordinator election and hence is highly scalable. Third, it provides nearly as much sensing coverage compared to the coverage obtained if all the nodes are awake. Finally, it preserves network connectivity by using a protocol based on CHECK and CHECK_REPLY messages. We next describe a basic scheme for self-configuration. The basic scheme will subsequently be extended to prevent network partitions.

4.3.1 Basic Scheme

In self-configuration based on SCARE, each node starts by generating a random number with uniform probability between 0 and 1. A node becomes eligible to be a coordinator if the random number thus generated is greater than a threshold (say 0.9). Therefore, a very small percentage of the nodes

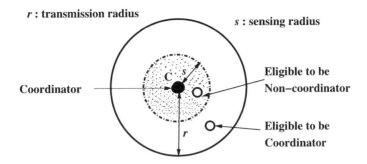

Fig. 4.1. Sensing and transmission radii of a node.

actually become coordinators. Other nodes just wait and listen. The threshold value can be preset depending on the application. A higher value for the threshold results in a small number of initial coordinator nodes. This has the effect of delaying the convergence of the self-configuration algorithm but it might result in a better selection of coordinator nodes. On the other hand, a low value for the threshold implies that a high number of coordinator nodes are selected randomly in the beginning. This hastens the convergence of the protocol although a larger number of coordinator nodes may be selected.

A node that is eligible to be a coordinator waits for a random amount of time before declaring itself to be a coordinator by broadcasting a HELLO message. This wait time, for example, can be chosen from a uniform distribution of values between T and NT where T is a preset slot time and N is the number of neighbors of the node that are coordinators. Initially N can be chosen to be a constant, e.g., 6. This prevents excessive contention on the wireless channel that might result if all the nodes decide to become coordinators at once.

Upon receipt of a HELLO message, a sensor node compares its distance from the sender C of the HELLO message to its sensing range s. A node within a distance s from a coordinator immediately becomes a non-coordinator node and stores the ID of the node that sent the HELLO message in its local cache. A node that is at a distance greater than s from C but within transmission range r becomes eligible to be a coordinator node. This is shown in the Fig. 4.1. The shaded region in the figure represents the sensing range of the node C. The outer circle represents the transmission range of the sensor node. Here, we assume that the sensing radius is smaller than the transmission radius. This is often the case for sensors in a sensor node [126].

While SCARE assumes the presence of an appropriate localization mechanism [5, 44, 9], exact distance calculations are not necessary. We show later that a moderate error in distance estimation has little effect on the outcome of the self-configuration procedure.

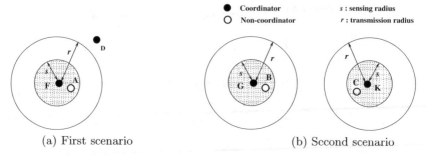

Fig. 4.2. Network partitions in the basic scheme.

4.3.2 Network Partitioning Problem

The basic scheme described above can sometimes result in a partitioning of the network; see Fig. 4.2(a). Here, coordinator node F makes node A a non-coordinator. However, coordinator node D can communicate with F only through A. This can potentially result in the partitioning of the network if coordinator (active) node D is unable to reach any other active nodes. As a result, network connectivity cannot be guaranteed in this situation. In Fig. 4.2(b), G and K are coordinator nodes and B and C are non-coordinator nodes. This situation again results in network partitioning as nodes G and K cannot reach each other. To prevent such situations, we extend the basic SCARE scheme outlined above, which results in a more effective technique for self-configuration. In the basic scheme, if there is a network partition, a node might never receive a HELLO message. This results in the node waiting eternally for a HELLO message, which results in wastage of energy. Hence, we choose a time-out value T_{off} after which the nodes that are still undecided about their state can become eligible to become coordinator nodes. The time-out value can be chosen based on the probability threshold discussed in Section III.A. A lower value for the threshold means that the procedure converges quickly and needs a lower T_{off} value, and vice-versa.

To prevent the network partitioning that occurs due to the pathological cases shown in Fig. 4.2, a node that initially receives a HELLO message from a coordinator node does not become a non-coordinator immediately and go to the sleep state. Instead, it continues to listen for messages from other coordinator nodes and remains in the "Eligible To be a Non-Coordinator" (ETNC) state. A sensor node that is in the ETNC state can become a coordinator node in two cases:

1. If it can connect two neighboring coordinator nodes[1] that cannot reach each other in one or two hops. It can deduce this information from the HELLO messages it received earlier. As shown in Fig. 4.2(a), node A,

---

[1] A neighboring node lies within the node's transmission radius.

which is in the ETNC state, receives HELLO messages from node F and node D, and decides to become a coordinator; this eliminates the partition.

2. If it can connect two neighboring coordinator nodes that cannot reach each other in one or two hops via a node in the ETNC state. As shown in Fig. 4.2(b), node B and node C, that are in the ETNC state, receive HELLO messages from node G and node K respectively, and decide to become coordinators as there is no match between the node lists of G and K; this eliminates the partition.

To achieve this, each ETNC node sends a CHECK message. This CHECK message contains the neighbor list of the coordinator node that caused this node to be in the ETNC state. Intuitively, this case is more likely to occur if there are few coordinators in the vicinity and less likely if there are more coordinator neighbors. Any ETNC node that receives this CHECK message replies with a CHECK_REPLY message and becomes a coordinator if there is no node common to the neighbor lists of both the nodes. Upon receipt of the CHECK_REPLY message, the node that sent the CHECK message also becomes a coordinator. In Fig. 4.2(b), non-coordinator node B sends a CHECK message and gets a CHECK_REPLY message from node C. Both node B and node C therefore become coordinator nodes. This procedure removes the network partition.

If the HELLO message is received from the same partition, and the node lists contained in the HELLO message do not have any common neighbors with the node lists the node received from other HELLO messages, then the node goes from the ETNC state to the "Eligible to be Coordinator" (ETC) state. This removes partitions if the HELLO messages are from different partitions. If they are from the same partitions, then the node connects the two coordinator nodes.

To prevent oscillations during the selection of coordinators, we enforce the condition that once a node becomes a coordinator, it continues to remain a coordinator until it is unable to provide any service. This strategy is used despite the fact that this coordinator might become redundant later during self-configuration. This penalty is reasonable since it occurs infrequently, especially in contrast to the energy needed to select an optimum number of coordinators. As the density of nodes increases, the fraction of non-coordinator nodes increases and this leads to more energy savings. Due to its distributed nature, SCARE a slightly larger number of coordinators than the minimum number necessary for coverage and connectivity. This also happens due to the randomness involved in the distributed selection of coordinator nodes.

After self-configuration, each coordinator periodically broadcasts a HELLO message along with a list of its one-hop neighbors that are coordinators. Non-coordinator nodes listen to these messages. Non-coordinator nodes also maintain a timer to keep track of the coordinator node that made them a non-coordinator. If this timer goes off, a non-coordinator node assumes that the corresponding coordinator node has failed and goes into an undecided state. This results in non-coordinator nodes becoming eligible to become coordinators.

SCARE can also be applied to mobile sensor networks. A node that has moved to a new location is treated in the same way as the appearance of a new node at that location. It sets itself to the Undecided state and listens to the network until either the timer T_{off} goes off or it receives a HELLO message. Similarly when a node moves away from one location, this is treated as a node failure by its neighbors. Failure of non-coordinator nodes does not result in any change in the topology. However, the movement of coordinator nodes is detected by the non-coordinator nodes and this makes them eligible to subsequently become coordinators.

4.4 Details of SCARE

A set of control rules governs the state of the sensor node while a set of defer rules decide when a node should postpone its decision. Timeout rules specify the time after which sensor nodes should make a decision.

A sensor node executing the SCARE procedure can be in one of the following states: Coordinator (C), Non-coordinator (NC), Eligible To be a Coordinator (ETC), Eligible To be a Non-Coordinator (ETNC), and Undecided (U). The ETC and ETNC states are temporary and exists only during the T_{setup} period explained below. There are seven timeout values in SCARE:

1. T_{off} : time after which a node that is in undecided state about its state becomes eligible to be a coordinator and goes into the ETC state.
2. T_{rand} : time for which the sensor node that is in ETC state waits before becoming a coordinator. It then sends a HELLO message along with all its coordinator neighbors that it has identified.
3. $T_{runtime}$: After every $T_{runtime}$ units of time, all non-coordinator nodes wake-up and listen.
4. T_{setup} : time interval for which the non-coordinator nodes wake up and listen, after which they go to sleep if they still remain non-coordinators. This is also the period during which beacon messages are sent to synchronize the nodes.
5. T_{coord} : time interval during which only the coordinators send HELLO messages. This occurs at the beginning of the T_{setup} period.
6. $T_{non-coord}$: time interval during which only the non-coordinators send messages. This is the latter part of the T_{setup} period. This period starts immediately after the T_{coord} period ends.
7. $T_{failure}$: A non-coordinator node waits for time $T_{failure}$ for the HELLO messages from the coordinator node that made it the non-coordinator. If no HELLO message is received within this time interval, it decides that the corresponding coordinator node has failed and sets its state to Undecided.

Next we describe the type of messages in more detail. There are three types of messages in SCARE:

- HELLO: These messages are sent by coordinators. They also contain a list of the one-hop coordinator neighbors of the sender node.
- CHECK: These messages are periodically sent by the non-coordinator nodes. They are used to remove the potential network partitions. Each CHECK message also contains of list of coordinator neighbors of the node that made it the non-coordinator.
- CHECK_REPLY: Upon receipt of a CHECK message, non-coordinator node compares the coordinator neighbor list included in the CHECK message with the neighbor list of the node that made it a non-coordinator. If there are no common entries in the two lists, it sends a CHECK_REPLY message. Thus, SCARE adopts a conservative strategy in creating paths in the network and prevent partitions. A non-coordinator node becomes a coordinator node if two coordinators at the end of the T_{coord} period cannot reach each other within one or two hops.

Recall that we used r to denote the transmission radius of a node. Similarly, recall that s is the sensing radius of a node. The control rules that decide the state of the sensor node are as follows:

1. A sensor node that generates a random number between 0 and 1, and greater than a threshold, becomes a coordinator.
2. A sensor node that lies at a distance between s and r of a coordinator node becomes eligible to become a coordinator node and goes into the ETC state.
3. A sensor node that lies at a distance at most s from a coordinator node becomes eligible to become a non-coordinator node and goes into the ETNC state.
4. A sensor node that is in ETNC state listens to the HELLO messages sent by the coordinator nodes for the T_{coord} period. From this list of coordinator nodes contained in the HELLO messages, if it determines that two coordinator nodes do not have a common neighbor that is a coordinator, this node becomes a coordinator at the end of the T_{coord} period. On the other hand, if there are common neighbors in the node lists, then the node stays in the ETNC state.
5. A sensor node that is in the ETNC state at the end of T_{coord} period broadcasts a CHECK message. This message contains a list of the coordinator neighbors of the node that caused it to go to the ETNC state.
6. A sensor node that receives a CHECK message compares the list of neighbors in the CHECK message with its neighbor list. If there is no match between the two lists, it transmits a CHECK_REPLY message to the sender of the CHECK message.

7. Upon receipt of a CHECK_REPLY reply to its CHECK message, a sender node that is in the ETNC state becomes a coordinator node. The node that sent the CHECK_REPLY also becomes a coordinator.

8. A sensor node that is in the ETNC state and does not satisfy conditions 4) and 5) becomes a non-coordinator node at the end of the setup period.

9. A sensor node that is in the ETC state becomes a coordinator node after the T_{coord} period if it does not become a non-coordinator node due to the selection of some other coordinator node.

10. A sensor node with data to send can opt to become a coordinator for as long as it has data to transmit.

The defer rules for SCARE are as follows:

1. If a node becomes eligible to be a coordinator, it listens for T_{rand} period.

2. If a node becomes eligible to be a non-coordinator at the end of the T_{coord} period, it listens for time $T_{non\text{-}coord}$ period.

The timeout rules are as follows:

1. A sensor node at the end of the T_{rand} period broadcasts a HELLO message.

2. A sensor node at the end of the T_{setup} period becomes a non-coordinator if it is still eligible to be a non-coordinator.

3. A sensor node at the end of the T_{coord} becomes a coordinator if it is still eligible to become a coordinator.

4. A sensor node wakes up and listens to the medium after the timer $T_{runtime}$ expires.

5. After its T_{off} timer expires, a sensor node becomes eligible to become a coordinator if it still undecided about its state.

A state diagram for the SCARE algorithm is shown in Fig. 4.3. The distance estimate is denoted by d, and we set $s = r/2$ in this figure. The time-out values in SCARE are application-dependent and they need to be tuned specific to the application. For example, the T_{off} value that triggers the state-transition from an *undecided* state to an ETNC state depends on the radio range of the specific sensor used in the sensor network.

4.4.1 Time Relationships

The relationships between T_{off}, $T_{runtime}$, T_{setup}, T_{coord} and $T_{non\text{-}coord}$ are as follows:

1. $T_{off} < T_{coord} < T_{setup}$.
2. $T_{coord} < T_{setup}$ and $T_{non\text{-}coord} < T_{setup}$.

These relationships are illustrated in Fig. 4.4.

Fig. 4.5 shows the result of applying SCARE to an example sensor network with 100 randomly deployed nodes in a 100m × 100m grid. The sensor nodes

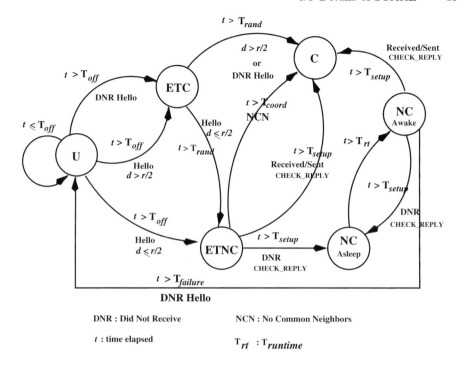

Fig. 4.3. The state diagram of SCARE.

have a radio range of 25m. Time-out values of $T_{failure}$ of 3s, T_{coord} of 3s, $T_{noncoord}$ of 2s, T_{setup} of 5s and $T_{runtime}$ of 95s are used. SCARE selects 32 nodes as coordinators and the rest are designated as non-coordinators.

4.4.2 Ensuring Network Connectivity

We next discuss how SCARE prevents network partitioning. Let S be a set of nodes containing the partial set of coordinators that are connected and the associated nodes in the ETNC state. Each coordinator in set S can reach any other coordinator in set S in a finite number of hops. Let X denote the region enclosing the nodes present in set S. Now consider a node not in set S. Any node not present in S can lead to the following scenarios. We use the notation P_A to represent the area within the transmission range of node A.

1. Coordinator B outside the region X but within the transmission range of the coordinator A in region X as shown in Fig. 4.6(a). In this case, both the coordinators can reach each other and the set $S = S \cup \{B\}$ and the region X expands to include the coordinator B.
2. Coordinator B is outside the transmission range of the coordinator A but is within the transmission range of ETNC node C; see Fig. 4.6(b). However, as node C listens to the HELLO messages from both coordinator nodes A

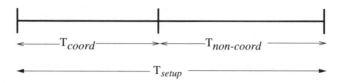

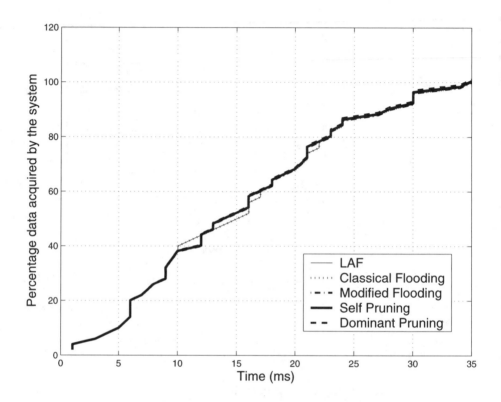

Fig. 4.4. Illustration of the relationships between the time intervals.

and B, it becomes a coordinator if there is no other path from A to B by becoming a coordinator. Now this reduces to (Case 1) with coordinators C and B within reach of each other. C becomes a coordinator and the region X expands to include the coordinator B, i.e. $S = S \cup \{B\}$.

3. Coordinator B is outside the transmission range of coordinator A. However, node C in ETNC state due to node B is within the reach of coordinator A as shown in Fig. 4.6(c). Node C listens to HELLO messages from A and B, and it becomes a coordinator. Now, A and C are within reach of each other and and this reduces to Case 2, hence $S = S \cup \{C\}$. By a similar procedure, node B is also included.

4. Coordinator B and coordinator A cannot reach each other as shown in Fig. 4.6(d). However, nodes C and D that are in ETNC state can reach

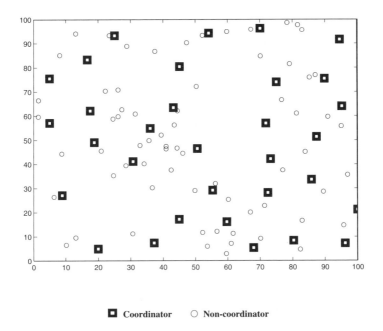

☐ Coordinator   ○ Non-coordinator

Fig. 4.5. Result of self-configuration using SCARE.

other. Node C and node D send and receive CHECK and CHECK_REPLY messages and become coordinators if there is no other path from node A to B. Once C becomes a coordinator, coordinator C in region X and co-ordinator D outside region X are within reach of each other. This reduces to case 2 and $S = S \cup \{D\}$. Region X expands to include node B and node D.

5. A node F that is outside the reach of either a coordinator or a node in ETNC state in region X. In this case, as the region X expands to include more nodes, node F falls into one of the above categories and as a consequence becomes connected with the nodes present in region X.

We have therefore shown that network partitioning can never arise during self-configuration.

4.4.3 Message Complexity

The total number of control messages, referred to as message complexity in SCARE, can be determined as follows: Suppose N is the total number of nodes in the network. Let N_c be the number of coordinator nodes selected. The number of non-coordinator nodes in the network is then simply $N - N_c$. Each co-ordinator node sends a HELLO message and each non-coordinator node sends a CHECK message. Let Δ be the average number of coordinator neighbors

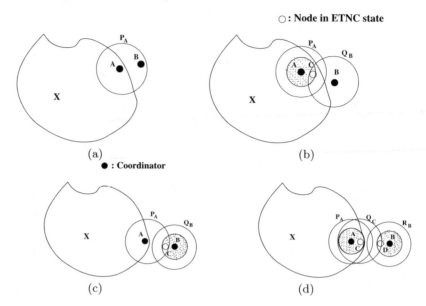

Fig. 4.6. Illustration of how network partitioning is prevented in SCARE.

of a non-coordinator node. A non-coordinator node sends a CHECK_REPLY message in response to a CHECK message if and only if there is no match between the coordinator neighbor lists of the non-coordinator nodes. In Span, each non-coordinator node sends one message and each coordinator node sends two messages. Therefore, the number of messages sent in each T_{period} interval is $N + N_c$.

Consider two non-coordinator nodes A and B. For every node in the coordinator neighbor list of A, let α be the probability that this node is present in the coordinator neighbor list of node B. The probability that there are is no match is then $(1 - \alpha)^{\Delta}$. Thus the expected number of CHECK_REPLY messages is $(1 - \alpha)^{\Delta} (N - N_c)$. The total expected number of control messages sent in SCARE is therefore $(1 - \alpha)^{\Delta} (N - N_c) + N_c + (N - N_c) \approx N$ for sufficiently large Δ in dense sensor networks. This is clearly less than the $N + N_c$ messages needed in Span. The size of each message in SCARE is almost equal to the size of each message in Span since the almost same information is contained in both sets of messages.

4.5 Optimal Centralized Algorithm

In this section, we develop a provably-optimal centralized algorithm that selects a smallest number of nodes to maximize coverage yet maintains network connectivity. This optimal algorithm is compared to SCARE in order to evaluate the effectiveness of the distributed algorithm.

We model the sensor network as a graph G, and use this model to develop an algorithm $MAXCOVSPAN$ that generates a spanning subgraph of G. In addition, G provides the maximum coverage among all spanning subgraps of G, where the nodes in the spanning subgraph correspond to the active sensor nodes in the network. The results provided by the centralized procedure $MAXCOVSPAN$ can then be compared with the results obtained using the distributed SCARE procedure.

Recall that SCARE selects nodes as coordinators on the basis of a distance metric. $MAXCOVSPAN$ also uses the distance between nodes to include nodes in a spanning subgraph such that the coverage is maximized.

Problem Statement:
Find a spanning subgraph of G that provides the maximum coverage. The vertices in G correspond to the sensor nodes. If two nodes are within radio range of each other, an edge is included in G between the corresponding vertices. The weight of this edge denotes the distance between the two sensor nodes.

Tha algorithm is described in terms of the following rules that are applied to G.

Initialization:
Rule 0: Color all vertices white.

Rule 1: Start with an arbitrary node. Call this node Current and Color it black.

Selection:
Rule 2: Pick an adjacent vertex that is connected by an edge to the Current vertex of maximum weight and color this node black. Color all other neighbors of the Current node gray. Call the vertex that has most recently been colored black as Current.

Rule 3: If the vertex belonging to the longest edge is already colored black, follow Rule 4 else repeat Rule 2.

Rule 4: If there are still white vertices, pick a gray vertex that has most white neighbor vertices and call it Current.

Termination:
Rule 5: Repeat Rules 2, 3, 4 till all the vertices are colored either black or gray.

Theorem 4.1. *The algorithm $MAXCOVSPAN$ runs in $O(n^2)$ time for a graph with n vertices.*

Proof: At each time instant, one vertex is colored either black or gray. There are n vertices in the graph. However, we need to check for the remaining gray nodes that have white nodes as their neighbors. This takes $O(n)$ time as we might have to check all the n nodes in the worst case. Hence, it takes a total of $O(n^2)$ time to complete the algorithm.   ∎

Theorem 4.2. *$MAXCOVSPAN$ always generates a spanning subgraph.*

Proof: It suffices to show that at the end of the algorithm, each node is colored either gray or black. This can be shown as follows. According to *Rule 4*, if a gray node has white vertices as neighbors, then it is colored black and all its neighbors except the neighboring vertex belonging to the longest edge are colored gray. A black node has all its neighbors colored black or gray according to *Rule* 2. This completes the proof of the theorem.   ■

Theorem 4.3. *The spanning subgraph G' generated by MAXCOVSPAN provides the highest coverage among all spanning subgraphs of G that have the same number of nodes as G'.*

Proof: In order to avoid case-by-case analysis, we prove this theorem using mathematical induction. Suppose $G = (V, E)$ is the graph corresponding to the sensor network. Let $P_i = (V_i, E_i)$ denote the partial (incomplete) spanning subgraph of size i generated by *MAXCOVSPAN*. Let $Cov(P_i)$ denote the coverage obtained with P_i. Consider the base case $P_1 = (v_1, \phi)$, where v_1 is any node selected at random. $Cov(P_1)$ is the maximum as all nodes have the same sensing range and the coverage provided by any node is the same. Next we assume that that the coverage of P_n is the maximum among all partial spanning subgraphs of size n. The coverage provided by a partial connected spanning subgraph of size $n + 1$ is given by $Cov(P_{n+1}) = Cov(P_n) \bigcup Cov(v_2)$ where v_2 is the node added to the partial spanning subgraph of size n. For $Cov(P_{n+1})$ to be maximum, v_2 needs to have minimum overlap with P_n. This is ensured by *MAXCOVSPAN*. The algorithm selects the node that is farthest from the partial spanning subgraph and this results in the coverage of the new selected node to have minimum overlap with the partial spanning subgraph of size n. From this observation, it follows that $Cov(P_{n+1})$ is maximum. Hence by the principle of mathematical induction, we have shown that *MAXCOVSPAN* generates a connected spanning subgraph that provides maximum coverage.   ■

4.5.1 Coverage Comparisons

Fig. 4.7 shows the variation of coverage with the total nuber of nodes for three scenarios: all nodes awake, *MAXCOVSPAN*, and SCARE. We assume that the nodes are placed dropped randomly on a 100m × 100m grid. We assume a sensing range of 12.5m and a transmission range of 25 m. We vary the number of nodes from 50 to 300, and the results are averaged over 100 runs. The results show that the distributed SCARE procedure performs nearly as well as the centralized *MAXCOVSPAN* procedure,and for a large rnumber of deployed nodes, both these nethods perform nearly as well as the scheme of keeping all nodes awake.

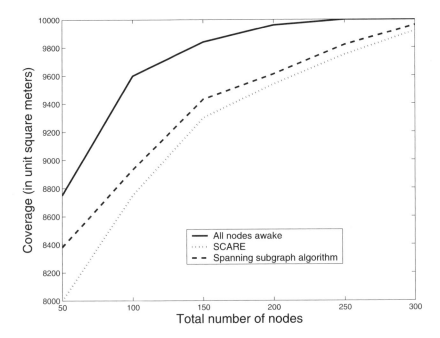

Fig. 4.7. Coverage versus number of nodes.

Table 4.1. Radio characteristics [6].

Radio Mode	Power Consumption (mW)
Transmit (T_x)	14.88
Receive (R_x)	12.50
Idle	12.36
Sleep	0.016

4.6 Performance Evaluation

To better understand the performance issues in SCARE, we use simulations to determine the effectiveness of SCARE in terms of coverage, connectedness, and network lifetime. We compare SCARE to Span in the simulations. Finally, we examine the impact of distance estimation errors on the effectiveness of SCARE.

Each sensor node is assumed to have a radio range of 25m. The bandwidth of the radio is assumed to be 20 Kbps. The sensor characteristics are given in Table 4.1 [6].

4.6.1 Simulation Methodology

We have developed a simulator in JAVA to evaluate the performance of SCARE. Our simulator uses geographic forwarding as the routing layer and IEEE 802.11 [127] as the MAC layer. Each sensor node that receives a packet forwards it to the neighbor coordinator node that is closest to the destination. If no neighboring coordinator node is closer to the destination than the node itself, the packet cannot be forwarded and it is dropped. SCARE runs on top of IEEE 802.11 MAC and below the routing layer to help coordinate the forwarding of packets from source to destination.

We use a grid size of 100 m × 100 m, and sensor nodes with radios having a nominal radio range of 25 m and a bandwidth of 20 Kbps. Initially, nodes are randomly deployed in the grid with the only condition that the nodes form a connected network. We simulate different node densities by increasing the number of nodes and keeping the grid size constant. To study the effect of increase in the number of nodes on SCARE, we simulate 50, 100, 150, 200, 250, and 300 nodes in our simulations. The results presented in this section are averaged over 100 simulation runs.

In the remainder of this section, we compare SCARE with Span and show that SCARE selects a smaller number of coordinators compared to Span and thus provides significant energy savings. To study the effect of SCARE coordinator selection on packet loss rate, we used a constant bit rate (CBR) traffic. However, to more closely understand the effectiveness of SCARE, we separate the nodes that generate traffic from the nodes that execute SCARE and participate in multi-hop forwarding of packets. Sources and destinations of traffic are placed outside the simulated region and these nodes do not execute the SCARE procedure. A total of ten source nodes and ten destination nodes are used in our simulations. Each source node selects a random destination from the 10 destination nodes and sends a CBR traffic of 10 Kbps to it.

To study the effect of mobility on SCARE, we use a random way-point model [17]. In this model, each node randomly selects a destination location in the sensor field and starts moving towards it with a velocity randomly chosen from a uniform distribution. Once it reaches the destination, it waits for a certain pre-determined pause time, before it starts moving again. The pause time determines the degree of mobility in this model. We simulated five different pause times of 0s 100 s, 200 s, 500 s, and 1000 s and a velocity range of 0 to 10 m/s. A pause time of 1000s corresponds to the stationary sensor network while a pause time of 0s corresponds to high mobility. We used $T_{failure} = 3$ s, $T_{coord} = 3$ s, $T_{noncoord} = 2$ s, $T_{setup} = 5$ s, and $T_{runtime} = 95$ s in our simulations.

Although SCARE relies on a localization scheme, for simplicity, we do not simulate it in our simulator. Instead, we make use of the geographic locations of sensor nodes provided by our simulator to aid SCARE in deciding the state of each sensor node. However, since the message overhead due to SCARE is negligible, only one messages per node, we believe that this does not affect the results significantly.

4.6.2 Simulation Results

In this subsection, we first evaluate the coverage provided by SCARE. We define coverage as the total sensing area spanned by all the coordinator nodes. We assume that non-coordinators nodes turn off their sensors. Although SCARE does not provide complete coverage due to the random deployment gaps in the sensing range of the coordinators, its coverage is very close to the maximum coverage. Yet, SCARE selects only a few nodes as coordinators to provide this coverage, thus achieving considerable energy savings. Therefore, SCARE efficiently trades off minimum loss in coverage with a tremendous gain in energy savings.

In Figure 4.8, we show the coverage versus the number of deployed nodes for SCARE. Recall that coverage is measured by the total sensing area spanned by all the coordinator nodes. We also show the coverage when SCARE is not run and all nodes are kept awake. As expected, the coverage obtained with SCARE is slightly less than the coverage obtained if all nodes have their sensors and radios turned on. However, the coverage produced by SCARE becomes comparable to the best-case coverage as the number of nodes increases. In these simulations, the grid size is kept constant, hence an increase in the number of nodes represents an increase in the node density.

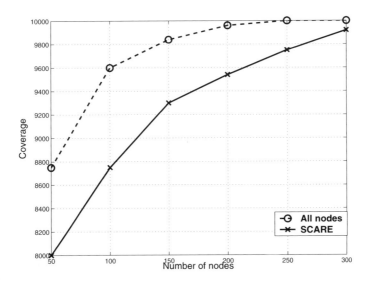

Fig. 4.8. Coverage obtained with SCARE compared to the case when all nodes are awake.

We next compare the number of coordinators selected in Span with the corresponding number for SCARE. As shown in Figure 4.9, the number of coordinators selected by SCARE is much less than in Span. (For 50 nodes, Span selects fewer coordinators, but the coverage is too low.) SCARE selects a smaller number of coordinators yet provides nearly the same coverage.

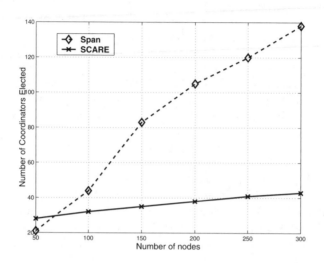

Fig. 4.9. Number of coordinators selected with an increase in nodes.

Figure 4.10 shows the coverage obtained by using SCARE and Span. SCARE tends to provide better coverage than Span for a range of values for the number of nodes below 100. Both provide similar coverage as the number of nodes increases beyond a threshold.

Figure 4.11 shows the fraction of nodes selected as coordinators with an increase in the number of nodes. SCARE selects a small fraction of nodes as coordinators with increase in node density. Hence, compared to Span, more energy savings are obtained with SCARE for dense sensor networks.

Figure 4.12 compares the number of coordinators selected by SCARE compared to the ideal number of coordinators needed for the square tiling configuration discussed in Section 4.5.1. SCARE selects almost the same number of coordinators as in the ideal case. This behavior is different from the behavior of SCARE in Figure 4.9 as here the nodes are placed in a regular fashion and not randomly deployed. Random deployment results in SCARE selecting more nodes as coordinators to cover the entire grid and still maintain connectivity.

Any self-configuration algorithm should have minimal control message overhead. In Figure 4.13, we compare the number of control messages used by SCARE and Span for the self-configuration. SCARE uses a smaller number of control messages compared to Span because it takes advantage of the random initialization of the nodes. This leads to a partial configuration of

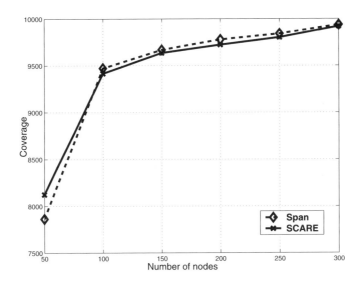

Fig. 4.10. Coverage versus number of nodes for SCARE and Span.

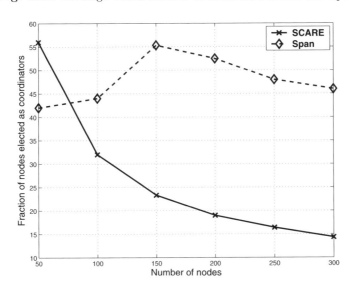

Fig. 4.11. Fraction of nodes selected as coordinators in SCARE and Span.

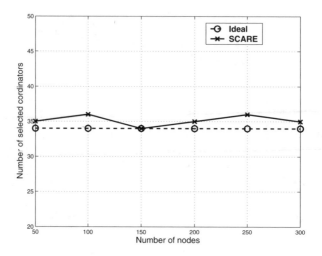

Fig. 4.12. Coordinators selected in SCARE versus an ideal number of coordinators selected based on square tiling.

the network, hence SCARE uses fewer number of control messages to achieve self-configuration.

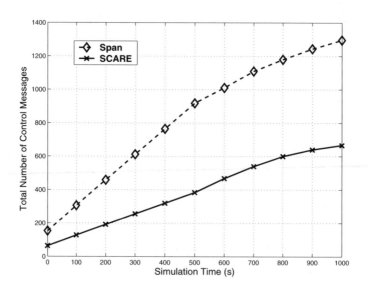

Fig. 4.13. Number of control messages used for self-configuration.

Figure 4.14 shows the effects of mobility on packet loss rate for both Span and SCARE. Nodes follow the random way-point model described in the previous subsection. Packet loss rate is calculated as the ratio of the number of lost packets to the number of packets actually sent. We note that the packet loss rates for both these methods are comparable.

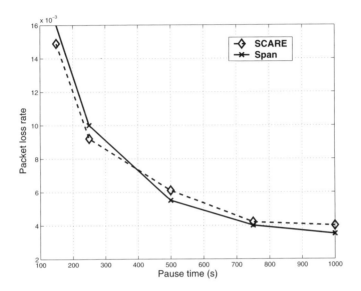

Fig. 4.14. Packet loss rate as a function of pause time.

Figure 4.15 shows the fraction of surviving nodes as a function of simulation time for both SCARE and Span. SCARE uses fewer control messages, and consumes less energy for self-configuration and re-configuration of the network. The number of surviving nodes falls below 80% at 765 s for SCARE compared to 700 s for Span.

4.6.3 Effect of Location Estimation Error

We next investigate how errors in distance estimation affect the performance of SCARE. Since nodes use distance estimation only to determine their eligibility to go to the sleep state, we do not expect SCARE to be significantly affected because of moderate errors in distance estimates.

To measure this feature of SCARE quantitatively, we ran simulations by introducing artificial errors in distance estimation. We modeled such errors by shifting the location of each node by a random amount in the range $[x \pm e, y \pm e]$, where e is either 10% or 20% of the radio range of a node and $[x, y]$ is the location of a sensor node. Nodes use these artificial locations rather than

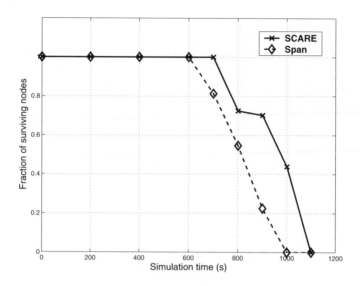

Fig. 4.15. Fraction of nodes remaining with time for Span and SCARE.

their real location to estimate the distance between them and a coordinator node. We refer to this scheme as either SCARE-10 or SCARE-20.

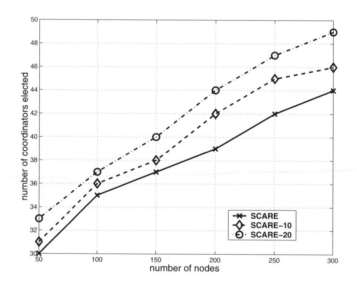

Fig. 4.16. Effect of error in distance estimation on SCARE.

Figure 4.16 shows the results of these simulations. The simulations using SCARE-10 and SCARE-20 are based on incorrect estimation of the distance from the coordinators by the nodes. Consequently, the number of coordinators is different from the case when there is no error. However, the increase in the number of coordinators is negligible while the decrease in coverage is found to be minimal. In the case of SCARE-10, the increase is only 3% for a small number of nodes and negligible ($< 0.2\%$) for a large number of nodes. The decrease in coverage was found to be at most 0.7%. As can be seen from the graph, the results are similar for the SCARE-20 case.

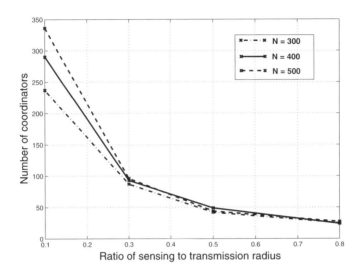

Fig. 4.17. Number of Coordinators selected versus s/r.

In all the simulations results shown above, the sensing radius has been taken to be one-half of the transmission radius. We now examine the affect of varying the sensing radius (s) as a fraction of the transmission radius (r) of a node. Figure 4.17 shows the number of coordinators selected by SCARE as the ratio of sensing radius to the transmission radius is varied. The number of coordinators selected drops rapidly as the ratio increases. As expected, the coverage increases with an increase in s/r; see Figure 4.18. At the s/r value of 0.3, we obtain almost 93% coverage with only 25% nodes selected as coordinators.

In the absence of calibrated data for the timeout parameters, we repeated the above set of experiments for different values of the parameters. The details are not listed here due to reasons of conciseness. We obtained similar experimental results in all cases.

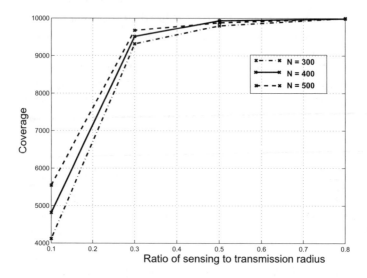

Fig. 4.18. Coverage obtained versus s/r.

4.7 Conclusion

In this chapter, we have presented a new scalable algorithm, termed SCARE, for self-configuration and adaptive reconfiguration in dense sensor networks. The proposed approach distributes the set of nodes into subsets of coordinator and non-coordinator nodes. It exploits the redundancy inherent in these networks to maintain coverage in the presence of node failure, as well as to prolong the network lifetime. We have presented a novel node replacement strategy that allows SCARE to use non-coordinator nodes to replace coordinator nodes that fail. We have presented simulation results to highlight the advantages of SCARE over a previously proposed topology management scheme Span for ad hoc networks.

5
Energy-Aware Information Dissemination

5.1 Introduction

As discussed in previous chapters, advances in the miniaturization of micro-electronic and mechanical structures (MEMS) have led to battery-powered sensor nodes that have sensing, communication and processing capabilities. Wireless sensor networks are networks of large numbers of such sensor nodes [18]. These networks are deployed in a target area for collaborative signal processing [99] to facilitate applications such as remote monitoring and surveillance [30, 95], tracking [19], and feature extraction [68]. Other applications of such sensor networks include the monitoring of wild fires, inventory tracking, assembly line monitoring, and networks of bio-sensors for health monitoring.

Sensor nodes are typically characterized by small form-factor, limited battery power, and a small amount of memory. Sensor networks enable distributed collection and processing of sensed data. These networks are usually connected to the outside world with base stations or access points through which a user can retrieve the sensed data for further inference and action. Base stations also send periodic control signals to all the sensor nodes. Such dissemination of information is a challenging problem in sensor networks because of resource constraints. For example, an intruder alert information might need to be forwarded to the entire network. Conventional protocols use classical flooding for disseminating data in a sensor network. Flooding is also used as a pre-processing step in many routing protocols in networks for disseminating route discovery requests [87]. Information dissemination protocols are used in networks for distributing the link state information. Routers in the Internet periodically use flooding to update link state information at other nodes [58]. However, flooding suffers from disadvantages such as the broadcast storm problem[14].

In this chapter[1], we present an energy-efficient flooding mechanism, termed location-aided flooding (LAF), for information dissemination in distributed sensor networks [108]. We have designed LAF with the following goals in mind:

- **Energy efficiency**: Sensor nodes have very small amount of battery and hence any solution must be energy-efficient.
- **Self-configuration**: Since it is not feasible to have manual intervention for every sensor node, it is preferred that nodes carry out self-configuration.
- **Scalable**: Sensor networks can typically have hundreds or thousands of nodes, hence any solution for information dissemination should be scalable.

We propose a solution that meets the above design requirements. We present an energy-efficient scheme that uses the concept of virtual grids to partition (self-configure) the set of nodes into groups of gateway nodes and internal nodes. While gateway nodes forward the packets across virtual grids, internal nodes forward the packets within a virtual grid. LAF reduces the number of redundant transmissions by storing a small amount of state information in a packet and inferring the information about nodes that already have the packet from the modified packet header.

Wireless sensor networks are different from ad hoc wireless networks in a number of ways, hence a data dissemination protocol for ad hoc networks does not immediately apply to sensor networks. An excellent survey highlighting the differences between ad hoc networks and sensor networks is presented in [3]. wireless sensor networks are used for obtaining sensing data from a monitoring area. Sensor nodes send data back to a base station that may be connected to the Internet, and where the data processing is done. This is typically not the case for wireless ad hoc networks. Ad hoc networks are typically used where there is no fixed infrastructure such as a base station. Ad hoc network routing protocols cannot be directly applied to sensor networks due to lack of scalability and the inability to adapt to a large number of sensor nodes [69]. These factors contribute to a dramatic increase in the control overhead for route discovery, topology updates, and neighbor discovery in ad hoc network routing protocols.

The remainder of the chapter is organized as follows. In the next section we discuss related prior work. In Section 3, we present the details of LAF. The performance evaluation of LAF is presented in Section 4. This section describes the simulation methodology, experimental results and a comparison with related work. Finally, conclusions and directions for future work are presented in Section 5.

---

[1] This chater is based on H. Sabbineni and K. Chakrabarty, "Location-aided flooding: An energy-efficient data dissemination protocol for wireless sensor networks", *IEEE Transactions on Computers*, vol. 54, pp. 36–46, January 2005.

5.2 Related Prior Work

In the classical flooding protocol, the source node starts by sending the packet that needs to be flooded to all of its neighbors [62]. Each recipient node stores a copy of the packet and rebroadcasts the packet exactly once. This process continues until all the nodes that are connected to the source have received the packet. This method of disseminating information is robust to node failures and it delivers the packet to all the nodes in the network provided the network is lossless. Flooding requires that nodes cache the source ID and the sequence number of the packet. This permits the nodes to uniquely identify each packet and prevents the broadcast of the same packet more than once.

A flooding algorithm based on neighborhood knowledge (self pruning), is presented in [75]. Each node obtains 1-hop neighbor information through periodic Hello packets. Each node includes a list of its one-hop neighbors in the header of each broadcast packet. A node receiving a broadcast packet compares its neighborlist to that of the sender's neighborlist. If the receiving node cannot reach any additional nodes, it does not broadcast the packet. The scalable broadcast algorithm presented in [92] uses 2-hop neighborhood information to limit the number of retransmissions. A node that receives a broadcast packet determines the 1-hop neighbors that need to rebroadcast the packet. A similar approach is taken in the dominant pruning method [128]. This approach uses the header 'trail' of the nodes recently visited by the packet to limit the number of broadcasts. It limits the length of the header trail by using a fixed hop-count. In contrast, LAF does not require that each node increase the length of the packet header; only a subset of nodes, referred to as internal nodes and defined later, are required to increase the length of the packet header. Furthermore, LAF uses gateway nodes (also defined in Section 3.5) to limit the header trail rather than using a fixed hop-count.

Information dissemination based on gossiping has been extensively studied in the literature [91, 89, 1]. In [1], gossiping is used to propagate updates among the nodes to maintain database consistency and it is used in [15] to provide reliable multicast. The performance of gossiping for wireless networks is compared with flooding in [49].

SPIN denotes a set of information dissemination protocols for wireless sensor networks [66]. In SPIN, nodes use meta-data to describe the data they possess. Nodes only request the part of data they do not have. Thus, SPIN achieves energy savings by eliminating requests for transmissions of data that nodes already possess. Although SPIN is also a flooding protocol, LAF is different from SPIN in two ways. First, LAF attempts to reduce redundant transmissions by inferring from the packet header about nodes that already have the data, while SPIN uses explicit communication to identify nodes that have the data. Second, while LAF uses location information to assist flooding and reduce energy, SPIN is a generic protocol that does not rely on location information.

Several approaches have been suggested to improve the efficiency of flooding using location information. LAR [114] uses the concept of a request zone to limit the search space for a desired route search. Location information has also been used in one of the five approaches suggested in [14] to contain the broadcast storm problem. A host node suppresses its transmission if the coverage provided by its transmission is less than a certain threshold. This coverage is determined from the locations of other nodes and by calculating the intersecting area of their transmission ranges. Multi-point relaying is proposed to reduce the number of re-transmissions due to flooding by choosing a set of relay nodes to broadcast the packets.

The concept of virtual grids in the context of routing is used in GAF [133]. All nodes in a virtual grid are equal from a routing perspective. GAF identifies redundant nodes within each virtual grid and switches off their radios to achieve energy savings. GAF cannot be used for flooding because of the small size of its grids; the density in the network has to be very high for nodes to take advantage of GAF for saving energy. GAF is mainly designed for routing where nodes in a virtual grid maintains the condition that at least one node in the virtual grid is awake. This results in a significant overhead if used for minimizing retransmissions in flooding. However, the goal of LAF is to limit the number of redundant transmissions during data dissemination in the sensor network, hence it differs significantly from GAF.

PAMAS [101] is a multi-access signaling protocol that conserves battery power by switching off nodes when they not actively transmitting or receiving packets. It uses a separate signaling channel for transmitting the control messages and for indicating a busy tone when a node is actively transmitting. The power savings are achieved because the signaling channel consumes less power compared to the main radio channel. Since LAF attempts to reduce redundant transmissions, this method can be used in conjunction with PAMAS to achieve higher energy savings.

Several solutions for the broadcast storm problem in flooding [14] have also been proposed. These approaches attempt to reduce the redundant broadcasts by allowing a node to suppress its transmission if some criterion is satisfied after receiving multiple copies of the flooded packet. LAF differs from [14] in a fundamental manner. It uses sender-based control to suppress the redundant transmissions rather than the receiver-based control used in [14] to reduce the redundant transmissions in classical flooding. A common feature of all these prior methods is that a node rebroadcasts a packet that all its neighbors have already received.

Finally, an efficient flooding mechanism based on passive clustering for on-demand routing routing protocols is presented in [69]. However, as described in [69], this method also suffers from scalability problems because even simulations are not feasible for networks with over 700 nodes due to excessive control overhead.

5.3 Location-Aided Flooding

The proposed protocol, which we describe in this section, uses a variant of classic flooding. We term this variant *modified flooding*. We describe the basic idea of modified flooding in the next subsection.

5.3.1 Modified Flooding

Modified flooding uses node ids to improve the energy efficiency of information dissemination in wireless sensor networks. Each packet sent using modified flooding includes a special field in the packet header called the Node List. Node List contains the ids of all the nodes that already have the packet. If we assume the network to be lossless, as is typically done in related literature, the packet header informs the receiver nodes that all the nodes in the Node List already have the packet, hence forwarding the packet to those nodes is unnecessary.

Modified flooding can be implemented in two ways. One option is to use a unicast scheme in which a sender node sends packets only to intended recipients. This is however difficult to implement in wireless networks. A more practical option is to allow all neighbors of the sender node to receive the packet through a broadcast mechanism. A recipient node first checks to see if all its neighbors are already in the Node List. If this test is affirmative, the node does not broadcast the packet. The node also checks to see if its own ID is in the Node List. If this test is affirmative, it does not process the packet and simply drops it. Note that the packet header (described later) contains source ID and sequence number information to facilitate modified flooding. In this work, we assume that the latter approach is used.

As an illustration, we show how the redundant transmissions in flooding can be reduced using modified flooding. When a node S wants to disseminate data to the entire network, it includes the ids of all of its neighbors[2] in the Node List of the packet header and broadcasts it to all its neighbors. Hereafter, we will refer to the node S as the *source* of the packet being flooded. A node, say X, after receiving the packet, retrieves the Node List of the packet and compares it with its neighbor list[3]. If all its neighbors are in the Node List, then X will not broadcast the packet. Thus redundant transmissions are avoided. Fig. 5.1 shows the operation of a modified flooding protocol for an example configuration. Node Lists are shown on the communication links. In this figure, node A wishes to flood the network with its sensor data. Therefore, node A broadcasts a packet with its data to all its neighbors. Nodes B and E also broadcast the packet, but there are no further broadcasts of this packet.

Although modified flooding results in energy savings by reducing redundant transmissions, the energy savings reduce as the Node List becomes

---

[2] Two nodes are neighbors if they are within the communication range of each other.

[3] Neighbor list of a node is the list of ids of all of its neighbors.

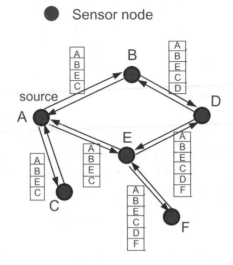

Classical flooding		Modified flooding	
Transmissions	Receptions	Transmissions	Receptions
A	B, C, E	A	B, C, E
B	A, D	B	A, D
C	A	E	A, F, D
D	B, E	C	A
E	A, D, F		
F	E		

Fig. 5.1. Example illustrating modified flooding.

longer; see Fig. 5.1. In classical flooding, each node broadcasts exactly once and every node receives all the broadcast packets of its neighbors. Hence, this simple network uses 6 transmissions and 12 receptions to flood the packet. Modified flooding uses 4 transmissions and 9 receptions to flood the same packet. However, if the packet becomes twice as long due to the increase in the length of Node List, this results in an effective number of 8 transmissions and 18 receptions. Thus, an increase in the Node List size limits the energy savings and the utility of modified flooding over classical flooding reduces beyond a certain point. To overcome this limitation, we describe our proposed approach, termed location-aided flooding (LAF), in the next section.

5.3.2 Location Information

LAF uses location information to divide the sensor network into virtual grids. This information may be provided by the Global Positioning System (GPS)

[56]. In GPS, receivers are used to estimate positions of the nodes in mobile ad-hoc networks. However, their high cost and the need for more precise location estimates make them unsuitable for sensor networks. GPS uses atomic clocks for time synchronization. Each GPS satellite transmits signals to sensor nodes on the ground indicating its location and current time. A node estimates the distance to each GPS satellite by estimating the amount of time it takes for the signal to reach the sensor node. Once the distance from four GPS satellites is estimated, a sensor node can calculate its position in three dimensions.

Several other localization schemes are also available in the literature for wireless sensor networks. In [90], a scheme is presented to estimate the relative location of nodes by having only a few nodes in the sensor network with GPS capability. It uses the received signal strength information (RSSI) as the ranging method to obtain accurate location estimates. [125] uses an ad-hoc localization technique called *Calamari* in combination with a calibration scheme to calculate distance between two nodes using a fusion of RF based RSSI and acoustic time of flight (TOF). Acoustic ranging [44] can be used to used to get fine-grained position estimates of nodes. [12] proposes a low-cost localization technique that uses time-of-arrival ranging. Recursive schemes such as [5] can also be used to get fine-grained position estimates of sensor nodes with error within 0.28m for nodes of 40m radio range.

For our initial discussion, we assume that each node knows its physical location accurately. However, we later show that LAF can tolerate moderate errors in location estimation as well as correlated large errors.

5.3.3 Virtual Grids

LAF divides the monitored area (sensor field) into "virtual grids". Each node associates itself with a virtual grid depending on its physical location. This is illustrated in Fig. 5.2, where the monitored area is divided into 9 virtual grids. Node A belongs to the virtual grid with the bottom-left-corner coordinates (2,2).

5.3.4 Packet Header Format

The header format of the packets used in LAF is shown in Fig. 5.3. It consists of the *SourceID* as well as the *SeqNumber* of the packet. The *recvNodeList* field is of variable length and contains the list of the nodes that have already received the packet. *GridID*, the ID of the grid in which the sender of the packet is currently in, and *nodeType*, whether the node is a gateway node or an internal node are the other fields. The field *gridID* is used only by the gateway nodes and is used for preventing the retransmission of packets the grid has already seen. The number of bytes for each field is best determined by the application designer. For example, the typical number of nodes in a sensor network application determines the number of bytes that needs to be reserved for the *gridID* field.

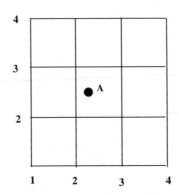

Fig. 5.2. Example of a virtual grid.

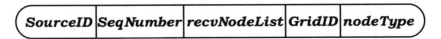

Fig. 5.3. Packet header format in LAF.

The size of a virtual grid and the appropriate number of virtual grids depend on the specific application requirements and also on the packet size. If the packet size is small, the overhead of the control data becomes significant after a few rounds of modified flooding. In this chapter, we assume that the number of virtual grids is determined *a priori*.

5.3.5 LAF Node Types

LAF classifies each sensor node into one of the two types:

- **Gateway Nodes**. If any of the neighbors of a node A belong to a different virtual grid than that of A, then A becomes a gateway node.
- **Internal Nodes**. If all the neighbors of a node A belong to the same virtual grid as that of A, then A becomes an internal node.

Nodes determine their virtual grid and status (gateway node or internal node) autonomously using the knowledge of their location information after deployment. This is the case for the example virtual grid shown in Fig. 5.4. Nodes A, G, F, I, D and H are gateway nodes while B, C and E are the internal nodes. Gateway nodes forward the data across virtual grids and internal nodes forward the data within a virtual grid.

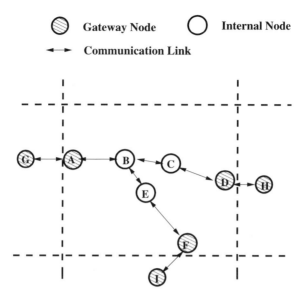

Fig. 5.4. Illustration of gateway nodes and internal nodes in a virtual grid.

5.3.6 Information Dissemination using LAF

Data forwarding by gateway nodes: When a gateway node receives a packet from within its virtual grid, it checks to see if any of its neighbors within the same virtual grid have not yet received the packet. This is done by comparing the Node List of the packet with the neighbor list of the node. If there exists such nodes, then the gateway node appends the ids of those nodes to the Node List of the packet and forwards it to the neighbor nodes that still have not received the message. When a gateway node receives a packet from another gateway node, it strips the packet of its Node List and adds its own id and all its neighbors' ids and forwards the packet to all its neighbors. Thus, the packet becomes shorter as it moves across the virtual grids and increases in size as it moves within a virtual grid. The basic idea behind LAF is to reduce the redundant transmissions by including the node ids in the packet. Virtual grids are used to limit the packet size. Gateway nodes in LAF cache the *sourceID* and *SeqNumber* fields of the recently seen packets. This allows the gateway nodes to prevent looping of packets in the network.

Data forwarding by internal nodes: When an internal node receives a packet, it modifies the Node List of the packet. It includes the ids of all its neighbors in the Node List and forwards it to its neighbors if they have not already received a message.

LAF is a simple protocol designed for lossless networks. However, it can be easily adapted for networks with error-prone communication links. Nodes can re-transmit a packet multiple times to compensate for the lossy communication links. Note also that LAF does not rely on a uniform placement of nodes. It can be easily used for a sensor network with "holes" because a regular virtual grid can be easily generated in a distributed manner for such a deployment.

5.3.7 Resource Management in LAF

LAF can be made resource-adaptive. When the remaining energy on the various nodes are different, nodes with less available energy can choose to wait for a time-out period before re-transmitting the packet that needs to be flooded. This time-out can be preset depending on the application requirements. The key idea behind this is that nodes with less remaining energy should participate only in the high priority tasks of the application leaving the low-priority tasks to the nodes with high remaining energy. (The alternative is to let all nodes participate to the same extent for all the tasks; however, this causes nodes with less remaining energy to die sooner.) This leads to a better utilization of the network for a longer period.

LAF does not specify a resource management policy and leaves it to the application to choose an appropriate policy depending on its latency and network lifetime and other application-oriented requirements.

Grid Maintenance Costs: Sensor nodes can use any unique attribute of their virtual grid as *gridID*. In our simulator, nodes use (x,y) as a *gridID* where x is the x-coordinate of the top left corner of the virtual grid and y is the y-coordinate. We estimate the cost of maintaining a grid as follows: Suppose the packet header size is h bits and the packet size, S bits. Also, suppose that the total number of n nodes are present in the monitored region divided into N grids. Suppose the number of beacon messages needed to know the position of a sensor node is n_B, and the energy needed to receive each beacon message is E_B. If each node needs n_p processing cycles to calculate grid association, then the total energy required to calculate the grid association is $n_p \times E_p$ where E_p denotes the amount of energy needed for a single processing cycle. Hence the total amount of energy needed for maintaining the grid is $n \times [n_B E_B + n_p E_p]$. Thus the grid maintenance cost grows only linearly with the size of the network.

5.3.8 Completeness of the Data Dissemination Procedure

In this section, we prove the completeness of LAF as a flooding mechanism. In other words, we show that data flooding can always be accomplished using LAF. A node that wants to flood the network with a data packet becomes the source for that data packet. We prove that if a node receives the data packet from the source through classical flooding, it will also receive the data packet through LAF.

Theorem 5.1. *If a gateway node in a virtual grid receives the packet, then all the nodes in the virtual grid will ultimately receive the packet provided that they received the message using the classical flooding protocol.*

Proof: Each node in a virtual grid can be either a gateway node or an internal node. Now, consider a node A in the virtual grid. Let us denote the neighborhood of A as the set N_A that consists of all the neighbors of A. Consider a node B in N_A that has the message. If the Node List received by B does not contain A, then B will forward the message to A. However, the Node List of the packet will contain the id of node A only if node A has the packet according to the LAF protocol. Thus, node A either has the message or it will receive the packet from node B. Once node A receives the packet it will forward it to all its neighbors that still have not received it. Thus, eventually all the nodes in the grid will receive the packet.   ■

Theorem 5.2. *If a source node floods the network with a message and if LAF is used by every node that forwards the message, then the message reaches every node in the network provided that the message reaches every node in classical flooding.*

Proof: We prove the theorem by contradiction. Consider a node in the random network that receives the message using flooding protocol but not using LAF. We call the node that is the originator of the message the source node and the node under consideration the destination node. Also we refer to the virtual grid in which the source node resides as the source virtual grid and the virtual grid in which the destination node resides as the destination virtual grid. Since the destination node has received the packet in flooding, there exists a path from the source to the destination. The destination node has not received the packet under LAF implies that none of the gateway nodes in the virtual grid of the destination have received the message (Theorem 1). If any of the gateway nodes received the message, they would have forwarded it to the destination node. This means that none of the neighboring virtual grids of the destination virtual grid received the message. If any of the neighboring grids received the message, they would have forwarded it to the gateway nodes of the destination virtual grid. By continuing in a similar fashion, we can show that the gateway nodes of the source virtual grid also did not receive the message. This implies that no message has been flooded in the source virtual grid, which is a contradiction. Hence, if each node in the network executes the LAF protocol, every node eventually will receive the flooded packet.   ■

Theorem 5.3. *: If node failures can only occur before flooding begins, the degree of fault tolerance of the network for LAF is the same as that for classical flooding.*

Proof: Suppose that some nodes in the network have failed *prior to* flooding. From Theorem 5.1, we know that if a message reaches a destination by classical flooding in the network with failed nodes, it will also reach the destination node by LAF. Thus the fault tolerance of LAF can be trivially shown to be equal to that of classical flooding.   ∎

An interesting open problem is to compare the fault tolerance of LAF with that of classical flooding if nodes fail during flooding. This problem needs to be investigated in more detail and is left for future work.

5.3.9 Analysis

In this section we first study two simple topologies and analyze the energy savings achieved by LAF compared to classical flooding. Then we derive equations for obtaining the energy savings due to LAF in random networks. Suppose the average size of a data message is S bits and the diameter of the network is D. (The diameter of a graph is the longest of the shortest paths between any two nodes.) If E_T is the amount of energy needed to transmit one bit of data and E_R is the amount of energy needed to receive one bit of data, the amount of energy consumed by a node sending the data message with k node ids and one of its neighbors receiving the message is $(S + ki)E_T + (S + ki)E_R$, where i denotes size of the node id in bits. For a network of N nodes with a fully-connected topology, for each packet that needs to be flooded, there are N transmissions and $N(N - 1)$ receptions. Therefore, the energy EC_{CF} consumed in the network is

$$EC_{CF} = S \times N \times E_T + S \times N(N - 1) \times E_R \qquad (5.1)$$

In LAF, since the message is transmitted with the node ids of all the nodes in the network, there will be one transmission and $(N - 1)$ receptions. If we ignore the small increase in packet length in LAF, the total energy EC_{LAF} consumed in flooding the packet is

$$EC_{LAF} = S \times E_T + S \times (N - 1) \times E_R \qquad (5.2)$$

For values of $N = 30$, $S = 64$ bytes, $E_T = 0.8 \ \mu$ J/bit and $E_R = 0.6 \ \mu$ J/bit, classical flooding consumes 280 mJ while LAF consumes approximately 9 mJ of energy.

As a second example, consider a line topology with N nodes. Each node has at most two neighbors. In this topology, for a message to be flooded, N transmissions and $2N - 2$ receptions are needed. This is due to the fact that in flooding, each node has to broadcast the packet exactly once and this results in N transmissions. As each node has to listen to all the transmissions of its neighbors, there are a total of $2N - 2$ receptions. Therefore the energy EC_{CF} consumed in flooding the message is

$$EC_{CF} = N \times S \times E_T + (2N - 2) \times S \times E_R \qquad (5.3)$$

In LAF, the message length is increased by one each time a node forwards the message in the line topology. A node will not process the message if it is included in the Node ID list. Hence there are only $N - 1$ transmissions and N receptions, the additional reception due to the Nth node, that receives the packet but need not transmit it as there are no neighboring nodes that have not yet received the packet. The total energy EC_{LAF} consumed in this case is

$$
\begin{aligned}
EC_{LAF} &= ((1 + 2 + \ldots + N - 1) \times i + (N - 1) \times S) \\
&\quad \times (E_T + E_R) \\
&= (N \times \frac{(N - 1)}{2} \times i + (N - 1) \times S)) \\
&\quad \times (E_T + E_R) + ((N - 1) \times i + S) \times E_R
\end{aligned}
\tag{5.4}
$$

For values of $N = 30$, $S = 64$ bytes, $E_T = 0.8\ \mu$ J/bit, $E_R = 0.6\ \mu$ J/bit and $i = 1$ byte, classical flooding consumes 12 mJ while LAF consumes approximately 6 mJ of energy.

Next, we analyze the energy savings in the case of a random network constructed as follows. Nodes are placed at random in a rectangular area. Nodes are battery-powered and have only a limited range for transmission. Two nodes are neighbors if they are within the transmission range r of each other. This type of random networks is useful for modeling a large number of practical situations involving ad hoc and sensor networks. Now, we derive equations that predict the energy savings for the LAF scheme. Consider a random network with a total number of N nodes, and with n nodes in each virtual grid. Suppose each node has Δ neighbors on average and M neighbors on average already have the packet. The increase in packet length due to addition of node ids is considered negligible in comparison to the total packet length. In LAF, a node does not process a packet if it is included in the Node ID list. The amount of energy consumed in flooding the virtual grid E_V using modified flooding is then given by

$$
E_V = [E_T + E_R \times M](n) \times S
\tag{5.5}
$$

Hence, the total energy EC_{LAF} consumed in flooding the packet throughout the network is

$$
EC_{LAF} = \frac{N}{n} \times E_V
\tag{5.6}
$$

In the case of classical flooding, the total energy EC_{CF} consumed is given by

$$
EC_{CF} = (E_T + E_R \times \Delta) \times N \times S
\tag{5.7}
$$

Fig. 5.5 compares the energy consumed by classical flooding with LAF with varying Δ and for different values of M, for $N = 100$, $n = 10$, $E_T = 0.8\mu$ J/bit, $E_R = 0.6\mu$ J/bit and $S = 64$ bytes.

5.3.10 Errors in Location Estimates

In the above discussion, we assumed that each node knows its geographical location precisely. However, there might be errors in the location estimate

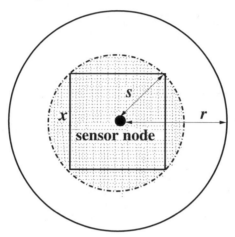

Fig. 5.5. Energy consumption for LAF (analytical result).

provided to the nodes by GPS[56] or other localization systems [90, 125, 5, 44]. Nevertheless, we do not expect the inaccuracies in position estimation to affect the performance of LAF. This is due to several reasons. First, LAF uses location information to associate a node with a specific grid. If the error in location estimate causes the node to assume a different location in the same grid, it will not affect the functioning of the node from a data dissemination viewpoint. Second, if the error in location estimate causes the node to assume a different virtual grid than the virtual grid it really belongs to, then the node becomes a gateway node in the assumed virtual grid and this also does not affect the performance of LAF significantly. Similarly, if a large correlated error causes a group of nodes belonging to a single virtual grid to be shifted to a different physical location, then the performance of LAF remains unaffected as all the nodes still belong to the same virtual grid.

5.4 Performance Evaluation

We have developed a simulator in C++ to evaluate the performance of LAF and compare it with alternative data dissemination algorithms. We found that LAF protocol achieves higher energy savings compared to both classical flooding and pruning-based methods while disseminating the data with comparable delay. We also found that the nodes with a higher degree (i.e., nodes with more one-hop neighbors) disseminate more data per unit energy in both LAF and modified flooding compared to classical flooding. Thus, dense sensor networks are likely to benefit more from using the LAF protocol for data dissemination in terms of energy savings.

5.4.1 Energy Model

Each sensor node is assumed to have a radio range of 20m. The bandwidth of the radio is assumed to be 20 Kbps. The sensor characteristics are given in Table 5.1. These values are taken from the specifications for the TR1000 radio from RF Monolithics [6].

Table 5.1. Radio characteristics [6].

Radio Mode	Power Consumption (mW)
Transmit (T_x)	14.88
Receive (R_x)	12.50
Idle	12.36
Sleep	0.016

5.4.2 Simulation Model

We initially used a 50-node network in a 200×200m monitoring area as shown in Fig. 5.6. The monitored area is divided into 4 virtual grids and has an average of 9 gateway nodes. This network is randomly generated with the pre-condition that the graph be completely connected. The processing delay for transmitting a packet is chosen randomly between 0ms and 5ms. This does not consider the queuing delays and other data processing delays that are incurred. We ran the data dissemination protocols 200 times and averaged the results. In each run, a randomly selected node floods the network with a 64-byte packet. Each node broadcasts a 5-byte HELLO packet every 2s. We have not implemented a localization system in our simulator. Instead, to compare energy consumption by LAF more accurately, we simulate the reception of three 10-byte beacon messages by the nodes every 2s. Finally, we assume that the network is lossless.

Although LAF relies on a localization scheme, we do not consider it in our simulator for simplicity. Instead, we make use of the geographic locations of sensor nodes provided by our simulator to determine the type of each sensor node (in practice, nodes determine their states autonomously). However, since the message overhead due to LAF is negligible, we believe that this does not affect the results significantly.

Data Acquired in the System with Time

Fig. 5.7 shows the percentage of data disseminated in the system with the passage of time for classical flooding, modified flooding, LAF, self-pruning [75] and dominant pruning [128]. As shown in the figure, the difference in message

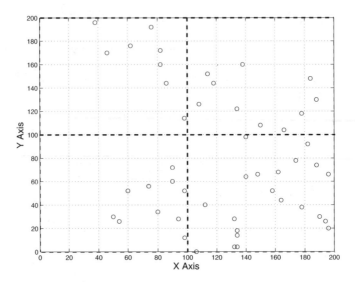

Fig. 5.6. Test network used in the simulations.

delay between these protocols is negligible. Fig. 5.8 shows a zoom-in view of Fig. 5.7. The difference in delay in these two protocols is visible in Fig. 5.7. This delay difference can be considered negligible for all practical purposes. The small difference in time delay arises due to an increase in message length in LAF and pruning-based methods and the corresponding increase in propagation time.

Energy Dissipated in the System with Time

Next, we measured the energy consumed in the system when these protocols are used for data dissemination purposes. Fig. 5.9 shows the total energy consumed in the system with time as data gets disseminated in the system. As shown, LAF achieves significant energy savings compared to the flooding protocol. (The energy consumption for LAF is less than 20mJ even after 35 ms.) By using a small amount of state information, LAF reduces the number of redundant transmissions significantly.

Impact of Number of Grids

We have varied the number of virtual grids for the test network shown in Fig. 5.6 and evaluated the performance of LAF using our simulator. Fig. 5.10 shows the energy consumption for the system when the monitored area is divided into 1, 4, 8 and 50 virtual grids respectively. The energy dissipated in the system decreases with an increase in the number of virtual grids up

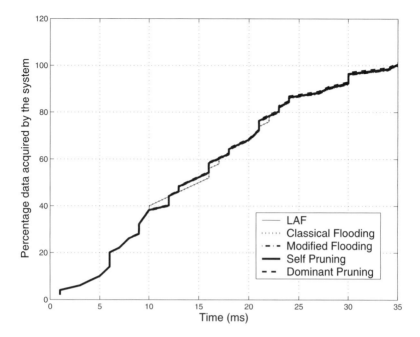

Fig. 5.7. Data disseminated in the system with time.

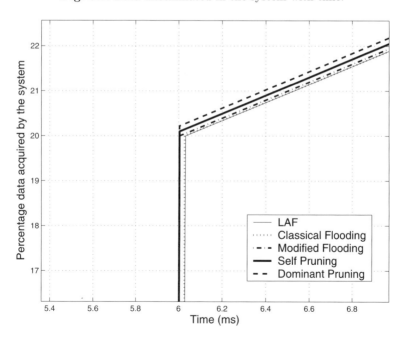

Fig. 5.8. A zoom-in view of Fig. 5.7.

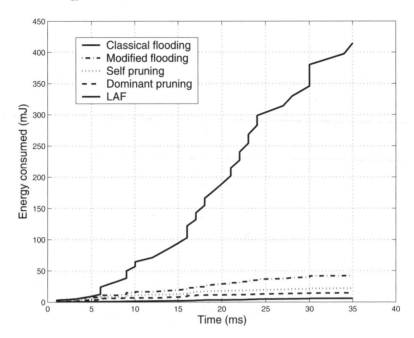

Fig. 5.9. Energy consumption for different data dissemination methods.

to a certain point, after which it decreases. This can be explained intuitively as follows. With a small number of virtual grids, energy savings due to the forwarding of the state information in the packet gets compensated by the increase in packet length. For a large number of virtual grids, the packet length remains within limits and the energy savings are significant. However, when the number of virtual grids is such that there are only a small number of sensor nodes in each virtual grid, the state information carried by the flooded packet within each virtual grid is very small and consequently the energy savings reduce.

Impact of Packet Size on Energy Savings

Typical packet sizes in a sensor network are 32 bytes, 64 bytes, 96 bytes and 128 bytes [129]. An increase in the size of the packet that is flooded results in an increase in energy savings. This is shown in Fig. 5.11 where three different packet sizes of 64 bytes, 96 bytes and 128 bytes respectively are shown. The energy savings are shown as the percentage savings in energy compared to the classical flooding protocol.

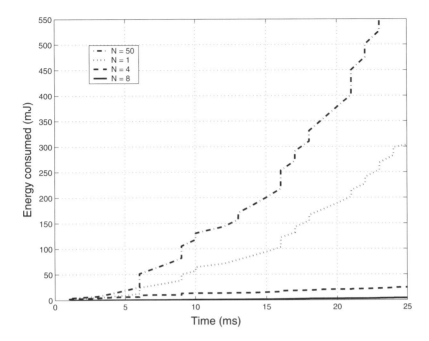

Fig. 5.10. Effect of number of virtual grids on energy consumption.

Impact of Degree of a node on Energy Savings

Fig. 5.12 shows the effect of the average degree of a node on the energy savings in LAF. A network of 100 nodes is divided into 8 grids and the energy consumed for the dissemination of a single 64-byte packet of data to 90%, 95% and 99%, of the nodes are plotted against the average degree of a node. The average degree of a node in the network is varied by changing the locations of the sensors. The total energy dissipated in the network decreases as the average degree of a node increases. This is because a larger number of redundant transmissions are avoided by making use of the information in the Node List.

Impact of Network Size on LAF

To study the scalability of network size on LAF, we varied the network size from 100 to 1000 nodes and flooded a single packet from a randomly selected source. Nodes were randomly deployed on a 200 × 200m grid and the entire grid was divided into 8 virtual grids. The results are shown in Fig. 5.13 and are averaged over 200 simulations. The figure shows that all of the methods are scalable except classical flooding. However, LAF outperforms all the existing methods in terms of energy savings.

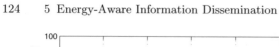

Fig. 5.11. Energy savings provided by LAF over flooding for different packet sizes.

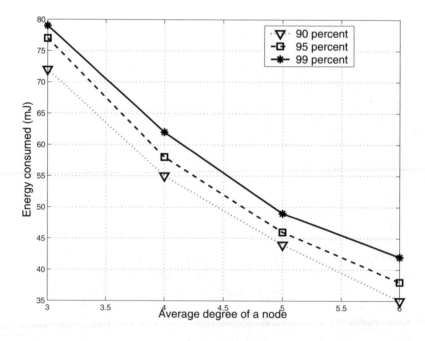

Fig. 5.12. Effect of average degree of a node on energy consumption.

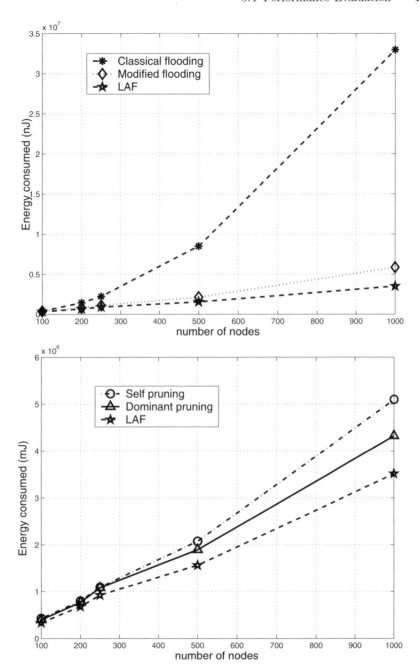

Fig. 5.13. Effect of network size on energy consumption for various data dissemination schemes.

Impact of Error in Location Estimate

To quantify the effect of error in location estimate on the performance of LAF, we repeated the above simulations by artificially introducing an error in the location estimates of the nodes. We introduced the error by shifting the location of each node by a random amount in the range $[x \pm e, \, y \pm e]$, where e is the error in the location estimate in terms of the percentage of the radio range of a node and $[x, y]$ is the actual location of a sensor node. Nodes use these artificial locations rather than their actual locations to associate themselves with the virtual grids and determine their type. We found in our simulations that up to 10% error in location estimate has negligible impact on the energy efficiency or the latency of LAF for data dissemination.

5.5 Conclusion

We have described a new energy-efficient flooding algorithm termed LAF for data dissemination in wireless sensor networks. The proposed approach uses the concept of virtual grids to divide the monitored area and nodes then self-assemble into groups of gateway nodes and internal nodes. It exploits the location information available to sensor nodes to prolong the lifetime of sensor network by reducing the redundant transmissions that are inherent in flooding.

This work can be extended by investigating the effect of using non-uniform grid sizes on the energy savings of LAF. Although in the above discussion we assumed a lossless network, LAF protocol can be easily adapted to lossy networks. A node can use the knowledge about the quality of a link to its neighbor and re-broadcast the packet multiple times.

Our results raise several interesting questions. First, we have used a uniform square grids in our simulations. However, a non-uniform grid size might be more desirable in situations where the node deployment is inherently non-uniform. Second, the size of the virtual grid can be tailored to the application and be adaptive to the activity in the network. Third, it is important to develop techniques that can dynamically reconfigure the virtual grid in a distributed manner after node failures, wearout and battery depletion. Finally, as part of future work, the energy savings need to be evaluated on physical hardware to demonstrate the usefulness of LAF.

6

Optimal Energy Equivalence Routing in Wireless Sensor Networks

The advance in MEMS, wireless communication, distributed computing, and sensor technology, has made wireless sensor networks (WSNs) flourish in recent years [3]. The untethered and unattended nature of WSN nodes destines most sensors to have energy sources which may not be replenished. Though some WSN units may have renewable energy such as solar battery, the expenditure will limit their application. Energy is a critical resource in wireless sensor networks. Energy efficiency and network longevity have dominated the design of WSNs and have occupied a large portion of research effort. Particularly in the research of routing protocols, the energy saving is an overwhelming consideration, usually combined with factors like delay and throughput.

Two paradigms are driving current research and development effort. The macrocosmic paradigm pays attention to network wide issues, especially network lifetime. From this point of view, energy saving is no longer an objective; rather it is a means to prolong the network lifetime. The microcosmic paradigm focuses on power consumption of sensor nodes, routing optimization.

When all sensor nodes have equal chances to become sensing sources and data sink, it is intuitive that the network will have maximum lifetime if all sensors consume energy at the same rate. In this chapter, we propose a new method to balance the energy consumption among all nodes. In other words, our paradigms keep approximate network wide energy equivalence.

This chapter describes the Energy Equivalence Routing (EER) methodology, a new approach developed at Louisiana State University following the macrocosmic viewpoint, which maintains network wide energy equivalence and maximizes network lifetime. Compared to other protocols, it emphasizes on route maintenance instead of route finding. In our approach, routes are not static any more. The quality of a route is varying while sensing, data transferring, and data aggregating is going on, especially for simultaneous multi-tasks scenarios, which are a normal situation for WSN.

For best fitting, routes are adapted periodically. No critical nodes would become the bottleneck of network lifetime. A reroute request packet is sent out from sinks periodically. When the packet reaches a path node, localized Common Neighbor Switching (CNS) algorithm checks energy difference between the node and its neighbors outside the routing tree. If the difference goes beyond a threshold, double neighbor switching is performed. Two path-rerouting algorithms, namely, Shortest Rerouting (EERS) and Longest Rerouting (EERL), are also presented to show that neighbor switching is better than path rerouting. Simulation results show that CNS outperforms Directed Diffusion, a typical existing protocol, in more than 90% cases, while EERS and EERL show only blurry and conditional advantage over directed diffusion.

To minimize energy cost the adjustment should be limited in as small a range as possible. Since this approach works in the route maintenance phase, it can be incorporated to any route finding protocols. Balancing is between nodes on routing trees and nodes outside routing trees, so working sensor nodes are replaced with unused nodes. In WSN, smallest topological range should be neighborhood, so our investigation concentrated on the relationship of neighbor.

This approach, as shown in simulation, is highly scalable in geographical size, however, it does not scale very well in network density. Neighbor switching needs a lower bound density, however, no upper bound is needed.

6.1 Related Work

6.1.1 Networking Characteristics of WSN

Wireless sensor networks fall in the category of wireless ad hoc networks. It could be mobile or fixed. They are divergent with conventional ad hoc networks, to make them work well we have to envision them in a brand new viewpoint and completely redesign them. Akyildiz et al particularized major differences between WSN and ad hoc networks in [3]. We made a little modification in the following list.

- The number of nodes in WSN may be several orders of magnitude higher than in ad hoc networks.
- Sensor nodes are prone to failure.
- Sensor nodes are limited in energy, computational capacities, and memory. So WSN can not afford table-driven MANET protocols requiring too much memory to store routing tables.
- Besides one-to-many broadcast model common in ad hoc networks, WSN often uses many-to-one communication model with the topology of the reverse multicast tree.
- Sensor nodes are densely deployed.
- The topology of WSN changes frequently.
- Sensor nodes may not have global (network wide or geographically global) ID.

Due to these differences, research focus was shifted to the search of special protocols for WSN.

6.1.2 WSN Protocol Stack

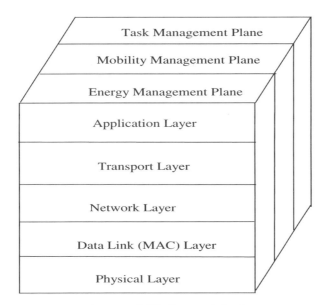

Fig. 6.1. WSN Protocol Stack

Imitating ISO OSI Model, Akyildiz et al. presented a protocol stack for WSN [3], which consists of five layers—Physical Layer, Data Link Layer, Network Layer, Transport Layer, Application Layer—and three planes, namely Energy Management Plane, Mobility Management Plane, and Task Management Plane, as shown in Fig. 6.1. In most cases, the stack is maintained in each sensor nodes; however, the proportion of implemented protocol stack is not homogeneous among nodes, usually, sinks, cluster heads, elected leaders would implement much more of the stack.

The physical layer deal with A/D, D/A transformation, modulation and demodulation, transceiver techniques like radio frequency carrying, optical transmission, right now CDMA is becoming mainstream in this layer. The data link layer, like in OSI Model, mainly concerns media access control (MAC) and error recovery. In radio WSN, frequency and bandwidth are very scarce due to the density of deployment. Channel or band sharing is ubiquitous, so MAC protocol is very important factor which affects the performance, efficiency, QoS, even final usability in WSN.

Routing is basic job in the network layer. One to many and many to one data transmission is one of fundamental features of WSN, so conventional point-to-point routing or end-to-end routing find much less use in WSN. Routing in WSN is always entangled with another principal trait of WSN, that is, precious energy supply. Almost no routing protocol till now has not address energy saving or its duality, i.e., prolonging network lifetime. In WSN, energy aware routing is the focus of focus; consequently major research effort converges on it, as our EER family.

In most wired or wireless ad-hoc networks, routing are based on a clumsy database or similar table-driven data structure, that is, routing table, which is built on almost every node and records and evaluates most possible paths through that node. Most routing algorithms input data from routing tables. However, WSN can not afford the cost of maintaining a routing table, which is against their adherent limits in memory, processing capability, power supply. Usually, a WSN adopts a much simpler routing data structure, just like a link list: every node, except sinks, only store current best one-hop next stop node, to the sink or to a source. This approach save a lot of resources, and obviously render a very special problem for routing in WSNs.

The transport layer is the interface of message and packet, like in ISO OSI Model. It provides ports or transport interfaces to various applications. The application layer decomposes application specific tasks and implements them using generic services offered by underlying layers.

The orthogonality between layers and planes shown in Fig. 6.1 has not been tested and is just for convenience. Similarly, the parallelism between planes is also an assumption or a hypothesis. We do doubt where the task management should go. It may also be conceptualized as a layer above or below the application layer. The independence of task management and energy management is probably a wrong assumption.

The energy management plane takes a network-wide control over energy consumption. It makes energy saving decisions strategies and general tactics. For instance, network-wide energy balancing, the basic idea behind EER family, is an output of the energy management plane. The mobility management plane detects and keeps track of the movement of sensor nodes. A sensor usually keeps a registry of their one hop neighbors, this information is sufficient if every node correctly registers the movement of its neighbors, which means backward routes are kept valid all the time. The task management plane schedules tasks and balance the tasks at the same time.

6.1.3 Classification of Energy Equivalence Routing

Generally, energy saving or energy aware routing protocols in WSN fall into two classifications. One is *architectural scale* of energy saving, which has two categories: sensor wide or network wide, like the macrocosmic paradigm and microcosmic paradigm we just discussed in previous section. Another is *routing scenario*, that is, route setup or route maintenance. EER runs in the route maintenance phase, it is dynamic and changes the routes many times. On the other hand, most other approaches work only in the route setting up, we call

Table 6.1. Where is EER

			Routing Scenario	
			Routes Setup	Route Maintenance
Energy Saving Scale	Sensor Wide			
	Network Wide	Data-Centric		EER
		Sensor-Centric		EER
		Address-Centric		

them static. EER is located in the intersection of sets network wide and route maintenance.

As shown in Table 6.1, network wide paradigm has three branches: *data-centric, sensor-centric*, and *address-centric*. Address centric routing is an old model, which has prevailed in traditional networks, like end-to-end routings in most traditional networks. Address-centric routing focuses on finding the shortest route between end-nodes via a global addressing mechanism. In WSN, however, data-centric approaches are preferred to address-centric routing to search routes from multiple sources to a single sink, to consolidate data from different sources, to avoid redundant transmissions, and to save energy eventually [73]. The key difference between data-centric paradigm and address-centric paradigm is that the former provides data aggregation. Sensor-centric paradigm, on the foundation of data-centric model, adds considerations about the failure of sensor nodes. It assumes sensors are autonomous entities but also organized in a team to serve a collaborative higher objective at the same time. In this way, sensors act as soldiers in an army, each one perish its own "life" but have to sacrifice themselves for the team if and only if they find the cost valuable and necessary. Section 2.5 will give a detailed comparison of EER to sensor-centric paradigm.

6.1.4 Energy Saving Routing Protocols

Many routing protocols have been designed to enhance the energy efficiency in WSN. At beginning on-demand protocols for mobile ad hoc networks (MANETs) were adopted for WSNs, for example, Ad Hoc On-Demand Distance Vector Routing (AODV). However, they acquitted themselves poorly, since they are inherently incompatible with WSNs in following aspects [3]. First, WSNs have stringent energy requirements. Second, WSNs have a much larger scale (node number) and a much higher density than normal MANETs. Third, WSNs often use many-to-one communication model with the topology of the reverse multicast tree. Furthermore, table-driven MANET protocols require too much memory to store routing tables, which WSNs can not afford. Finally research focus was shifted to the search of special protocols for WSN.

Many protocols devoted to WSN were developed. Among them, Flooding family, Small Minimum Energy Communication Network (SMECN) [115], Low Energy Adaptive Clustering Hierarchy (LEACH) [105], and Sequential Assignment Routing (SAR) [67] represent the mainstream in research. Since Flooding family, especially Directed Diffusion, is tightly related to EER, we give a detailed introduction in Sections 2.3 and 2.5 and skip it in this section.

Based on MECN [74], which was actually designed for ad hoc networks but works well in WSNs, SMECN is more efficient than MECN. Both SMECN and MECN derive an energy efficient subnetwork from a given network; however, SMECN gives a smaller subnetwork. The subnetwork contains all nodes in the original network, but it contains only a subset of original edges set. For any node, energy for broadcasting to all its neighbors in the subnetwork is less than the energy for broadcasting to all its neighbors in the original network. Any two nodes must be connected in the subnetwork if they are connected in the original network. Furthermore, a minimum energy path exists in the subnetwork between any two nodes connected in the original network.

SAR builds a tree for every neighbor of sink. Avoiding low energy residual nodes and low QoS nodes, trees grow outward from sink until all nodes has been traversed. Eventually most nodes are covered in one tree. When passing data to sink, each source selects appropriate path in tree based on the energy resource and QoS metrics of trees. SAR uses single winner election algorithm and multiple winner election algorithm to implement data transfer and synchronization. QoS metrics include delay, throughput, etc.

In the conceptual framework, LEACH is most similar to EER among known protocols. While investigating EER, we find that static routing is not optimal for WSNs. Similarly, static clustering is not optimal for WSNs either. LEACH randomly rotate cluster heads to evenly distribute the energy dissipation among sensors in the network. Like EER LEACH remarkably prolongs network lifetime. Localized coordination enables LEACH to enjoy scalability and robustness. LEACH also incorporates data fusion to reduce the amount of data that must be transmitted. LEACH achieved up to 8 times reduction in energy dissipation compared with previous routing protocols.

LEACH saves energy by clustering, such that nodes go through a tree like hierarchy to communicate with sink. LEACH randomly selects cluster head. Like our EER, LEACH has two phases in routing: route setup phase and steady phase. In setup phase cluster heads is selected by comparing a random number with a threshold from unselected nodes.

$$threshold = \frac{P}{1 - P[r \bmod (1/P)}$$

P is adjustable percentage of cluster heads, r is the round number. These cluster heads broadcast to other nodes that they has become cluster heads. Other nodes then select which cluster it would like to join. Unlike route maintenance phase in EER family, steady phase in LEACH does nothing about routing.

The only job is sensing and data transferring. Cluster heads also aggregate data from cluster members.

6.1.5 Comparison to Flooding Family

Flooding is a natural way for multi-hop communication, but it consumes too much resource, especially energy. Many variants of flooding have been presented [16], many of them were proposed by the research group in UCLA led by Deborah Estrin. For example, GRAdient Broadcast (GRAB) [130] sets up cost field by flooding. Gossip [91] exercises a partial probabilistic flooding to diffuse interests or events. Sensor Protocols for Information via Negotiation (SPIN) [66] use metadata message in the lieu of sensed data to reduce the amount of data required to transfer. The well-known Directed Diffusion [129] protocol uses limited flooding and an acknowledgement scheme to set up route between sources and sinks. The scheme is integrated with data aggregation to minimize the communication and maximize the energy efficiency, so it is data-centric. Geographical and Energy Aware Routing (GEAR) [131] bounds the flooding to a small geographical region. The recent Rumor protocol not only integrates Gossip and GRAB into directed diffusion but also combines query flooding and event flooding. It reaches a better performance [16]. A detailed comparison of these routing protocols and our approaches is given in Tables 6.2 and 6.3.

GRAB first builds up a cost field toward sink, then direct queries toward that node across a relatively small mesh. A cost field is similar to the gravity field that drives water flowing from a high post to a low post. A message flows to sink like water. The cost field value at each node is the minimum cost to reach the sink from that node. Once the cost field is established, any sensor can deliver the data to the sink along the minimum cost path. The cost field is created at the price of network flooding, but following data and queries are routed along shortest paths, and can thus be delivered cheaply and reliably. Each intermediate node forwards the message only if the consumed cost plus the cost at this node is equal to the cost of source. Therefore minimum cost path is realized without maintaining explicit path information. Like EER, GRAB adapts the established routes to the variation. GRAB employs a hybrid approach of event driven and timeout to refresh the cost field. The sink monitors certain properties of received packets such as success ratio and hop numbers. Once there is a major change, it rebroadcasts a message to rebuild the cost field. If no report is received for certain time, the sink also refreshes the cost field upon such a timeout.

GRAB allows a packet to carry a cost credit beyond the minimum required, then the packet can go to sink through different paths which form a mesh. By assigning different amount of credit, GRAB adjusts the width of the mesh and consequently the redundancy and robustness of the transferring. GRAB uses a simple algorithm that linearly spread the credit from the source to the sink with a high delivery ratio even in cases of severe environment noise and node failures.

Gossip provides a scheme for performing reliable network broadcasts, probabilistically. In Gossip protocol nodes send a message to some of the neighbors, usually only one, instead of all. The recipient is usually selected randomly, but some variants use deterministic algorithms for selection. [16] Due to the redundancy in the links, most nodes received the packet in limited steps (hops), so Gossip has similar effect like flooding. This scheme can be used to deliver queries or events for gradient setup, with less overhead than traditional flooding. Although Gossip routing was not initially designed for energy saving, but it reduce energy consumption by minimizing amount of transportation. Some other advantages are being investigated.

Unlike previous routing protocols in flooding family, Rumor protocol [63] tries to combine query flooding and event flooding. As shown in Fig. 6.2, Rumor first creates paths leading to each event. Event flooding sets up network-wide gradient field, which is made up of event paths. When a query is generated later, instead of being flooded across the network it is sent on a random walk until it runs into an event path.

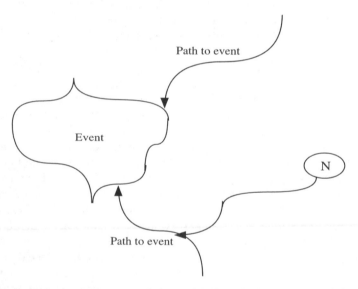

Fig. 6.2. A query searches event path instead sink with greater chance of success [2].

In sensor networks, queries and events are the only starting points of a sensing task cycle. They are opposite and symmetric in semantics. Queries are flooded from a sink to sources, whereas events need to be advertised from source to sink. A query is a request to specific data. A query is often described as a particular interest for data, which consists of a set of requirements for sensors and for data, so sometimes it is called an interest. Once the query

Table 6.2. Comparison of Previous Protocols and Our Approaches

Properties Protocols	Authors	Flooding	Neighbor Switching	Rerouting
Previous Approaches				
Generic Flooding		Query or event, many conventional flooding (unlimited, for each data sensed)	No	No
GRAB	Fan Ye et al (UCLA)	Query, 1 conventional to set up path	No	No
Gossip	Meng-Jang Lin et al (UCSD)	Query or events, probabilistically limited	No	No
Directed Diffusion	Chalermek Intanagonwiwat et al (USC)	Query, 1 conventional to set up paths	No	No
GEAR	Yan Yu et al (UCLA)	Geographically limited	No	No
Rumor	David Braginsky et al (UCLA)	Both query and event, both are limited, queries are random	No	No
Our Approaches				
EERS	Wei Ding, S.S. Iyengar, Rajgopal Kannan (LSU)	1 conventional to set up paths, followed by many short and limited	Partial and limited (1 node)	Short
EERL		1 conventional to set up paths, followed by many short and limited, but much fewer than EERS	Partial and limited (1 node)	Long
CNS-Plain		1 conventional to set up paths	Common neighbor (3 nodes)	No
CNS-Alternative		1 conventional to set up paths	Common neighbor (3 nodes)	No
CNS-Intermediate		1 conventional to set up paths	Common neighbor (3 nodes)	No

Table 6.3. Comparison of Previous Flooding Protocols and Our Approaches (Continued).

Properties Protocols	Tree Structure	Energy Saving Scheme	Lifetime	Wake up Mechanism
Previous Approaches				
Flooding	No when flooding	No	Shortest	
GRAB	No when flooding, no after flooding (mesh)	By limiting flooding and use of paths	Not compared	No, but could be added
Gossip	No when flooding	By limiting flooding	Not compared	No, but could be added
Directed Diffusion	No when flooding, yes in reinforcement	By limiting flooding and use of paths	Fair	No, but could be added
GEAR	No when flooding, yes in reinforcement	By limiting flooding	Not compared	No, but could be added
Rumor	No when flooding, yes in reinforcement		Not compared	No, but could be added
Our Approaches				
EERS	No when flooding, yes in reinforcement	By limiting flooding, use of paths, and energy balancing (at the price of additional flooding)	Sometimes better than directed diffusion	limited
EERL	No when flooding, yes in reinforcement	By limiting flooding, use of paths, and energy balancing (at the price of additional flooding, less than EERS)	Sometimes better than directed diffusion	limited
CNS-Plain	No when flooding, yes in reinforcement	By limiting flooding, use of paths, and energy balancing	Long	Thoroughly used
CNS-Alternative	No when flooding, yes in reinforcement	By limiting flooding, use of paths, and enhanced energy balancing	Long, could be better than CNS	Thoroughly used
CNS-Intermediate	No when flooding, yes in reinforcement	By limiting flooding, use of paths, and enhanced energy balancing	Long, could be better than CNS	Thoroughly used

arrives at its destination, referred to as a source, data begins to flow back to the sink, the query's originator. An event is something to be sensed or to be transmitted, ranging from a sensor reading to the energy residual of a node.

In most protocols in flooding family, queries are broadcasted. However, it is more economical and efficient to discover short paths from the source to the sink if the amount of returning data is significant. When there are few events and many queries, event should be flooded instead of query gradients towards it. As soon as the query discovers the event path, it can be routed directly to the event. If the path cannot be found, the application can resubmit the query, or as a last resort, flood it. The underlying heuristic is very simple: two straight lines in a plane are very likely to intersect. Although neither the path nor the query is perfectly straight, their intersection is still very likely.

GEAR relies on localized nodes, and provides savings over a complete network flood by limiting the flooding to a geographical region.

Directed Diffusion results in high quality paths, but requires an initial flood of the query. One of its primary contributions is an architecture that names data and that is intended to support in network processing. We will introduce this paradigm in Section 6.2.7 with much more detail.

6.1.6 Comparison to Sensor-Centric Paradigm

Basically the sensor-centric approach extends the data-centric routing algorithms. Data-centric approaches search for the most reliable paths and finally construct a most reliable tree. Sensor-centric approach modifies the most reliable tree or path by deviating the most reliable tree at data aggregation node according to the new consideration of node life, node cost and energy consumption.

In sensor-centric paradigm, sensors have more autonomy. They may decide whether to participate in routing; they may consider maximizing their individual lifetime as a criterion. The paradigm adds two more realistic constraints to data–centric paradigm: (1) possibility of sensor failure; (2) sensors must cooperate to achieve network wide objective while maximizing their individual lifetime. The sensor-centric paradigm uses a game-theoretic routing model in which rational sensors select routing paths by evaluating the trade-offs between reliability and the costs of communication. In sensor-centric paradigm, a game theory based routing heuristic, which is called *reliable query reporting*, is adopted.

The paradigm uses *quality of routing* (QoR) concept for evaluating data-aggregated routing trees, the metrics of which is called *path weakness*. It estimates how much a node would gain by deviating from the optimal data-aggregated tree. Furthermore, the paradigm can evaluate other routing algorithms on the complexity of paths with bounded weakness. The paradigm is a very comprehensive model and can be used in all phases of WSN routing — modeling, analysis, evaluation and algorithm design.

First, like the principal difference with flooding family, EER basically differs with sensor-centric paradigm in the time to take effect. EER always works in the phase of route maintenance, while most other protocols, data-centric or sensor-centric, work in the route setup phase. Second, although some EER approach like CNS could be directly used in failure recovery, EER has not included considerations about node failure, whereas sensor-centric paradigm pays much attention to sensor failure. Actually sensor-centric paradigm uses node failure probability everywhere in its quantitative analysis. Third, nodes in EER do not enjoy much autonomy, they are not allowed to freely join or leave, while in sensor-centric paradigm each sensor has sufficient authority to use its own judgement to decide enter or exit.

6.1.7 Data-Centric Routing and Directed Diffusion

In MANETs and other traditional networks, *address-centric* routing focuses on searching the shortest route between end-nodes via a global addressing mechanism. In WSNs, however, *data-centric* approaches are proposed in lieu of address-centric routing to search routes from multiple sources to a single sink, to consolidate data from different sources, to avoid redundant transmissions and to save energy [73].

Directed diffusion is a *data-centric* routes setup paradigm to efficiently disseminate sensed data in WSNs. In Directed diffusion data is named using *attribute-value pairs* without network-wide addressing. A sink disseminates a task as an *interest* for named data throughout the network. This flooding style dissemination establishes *gradients* to draw matching data. Sensed data is an *event* which is disseminated in reverse directions of interests. Events start towards the sink along multiple paths. The sensor network reinforces one, or a small number of these paths.

In directed diffusion, task descriptions are named by a list of attribute-value pairs. A task description specifies an interest for data matching the attributes. The data sent in response to interests are also named using a similar scheme. For each active task, the sink periodically broadcasts an interest message to each of its neighbors. This initial interest contains a much larger interval attribute. This initial interest is exploratory, trying to determine if there are any sensor nodes that detect the target. So the initial interest also species a low data rate. Below is an attribute-value description of a tracking task.

    type = tracking
    target = tank
    interval = 0.05 seconds
    duration = 100 seconds
    timestamp = 02:20:40
    expiresAt = 02:35:00
    rect = [200, -100, 300, 400]

Attribute **type** specifies classification of the task. Attribute **target** specifies the sort of target to be tracked. Attribute **interval** specifies the sensing interval for potential source nodes. Attribute **duration** specifies total duration of sensing. the task state is purged from passed nodes after the time. Attribute **timestamp** specifies the sent time of the interest in the format of hh:mm:ss. Attribute **expiresAt** specifies the ending time of the task, also in the format of hh:mm:ss. Attribute **rect** specifies rectangular area of target, which defines sources.

The interest is periodically refreshed by the sink via resending the same interest with a monotonically increasing timestamp, for interests are not reliably transmitted throughout the network. Every node maintains an interest cache. Two interests in cache are regarded distinct if their type, target, interval, or rect differs. Interest entries do not contain information about the sink. Thus, interest state scales with the number of distinct active interests.

An entry in the interest cache has several fields. A timestamp field indicates the arrival time of last received matching interest. The interest entry also contains several gradient fields, each for one neighbor. The data rate gradient field, which derived from the interval attribute of the interest, is requested by the specified neighbor. The duration gradient field, derived from the timestamp and expiresAt attributes of the interest, specifies the lifetime of the interest.

Having received an interest, a node checks if it exists in the cache. If no matching found, the node creates an interest entry. The parameters of the interest entry are instantiated from the received interest. This entry has a single gradient towards the neighbor from which the interest was received. If an interest entry has no gradient for the sender of the interest, the node adds a gradient with the specified value. It also updates the entry's timestamp and duration fields. If both an entry and a gradient exist, the node simply updates the timestamp and duration fields.

Sink initially diffuses an interest with a low event-rate. Once sources detect a matching target, they floods low-rate events to the sink. After the sink starts receiving these low data rate events, it reinforces one particular neighbor in order to draw down higher data rate events. One method could be to reinforce any neighbor from which a node receives a previously unseen event. To reinforce this neighbor, the sink resends the original interest with a smaller interval. When the neighboring node receives this interest via a gradient, it finds a higher data rate than before. If this new data rate is also higher than that of any existing gradient, the node must reinforce at least one neighbor as well. Using its data cache, this node might choose that neighbor from whom it first received the latest event matching the interest. Alternatively, it might choose all neighbors from which new events were recently received. Through this sequence, a path is established from source to sink transmission for high data rate events.

6.2 Energy Equivalence Approach

6.2.1 Basics

The energy equivalence approach is motivated by the observation that network longevity and energy efficiency have different semantics. Usually energy saving is associated with individual sensor nodes while prolonging the lifetime of network is a network-wide issue. If energy saving is disseminated equally among nodes, network lifetime would be maximized. However, if the saving is distributed unevenly, especially if the high energy consuming subgraph is critical to the overall connectivity, the unevenness is disastrous to the lifetime of network. We call the critical subgraph the *topological bottleneck* of the network. As a result, there may be some networks in which nodes are not very energy-efficient but networks as a whole achieve optimal lifetime. In this chapter we explore this phenomenon and discuss our observations and discoveries.

Typical topological bottlenecks include aggregating nodes and the common subnets on which several concurrent tasks overlap. Fig. 6.3 shows an example of aggregating node. In the example, we assume the radio radius is not long, so listening and transmitting dissipate same energy [7]. One by one five sensing nodes transmit data to the aggregator at interval of 10 time units. Each 10 time units, every sensing node dissipates 1 unit energy. The aggregator receives 5 data packets, integrates them into one packet and then passes it to sink, so for each relay, the aggregator dissipates 6 unit energy. After 8 relays, the aggregator has lost $6 \times 8 = 48$ unit energy. So its energy residue is 2, while energy residue of other nodes is 42.

A *routing tree* consists of the reinforced paths like that in Directed Diffusion. In data-centric routing paradigms like Directed Diffusion, two or more branches of a routing tree converge at an aggregation node, which is called aggregator in some papers (see Fig. 6.4). In this chapter the term *converging node* is used instead.

In this chapter, we propose a method called neighbor switching to balance the energy consumption among all nodes and remove the topological bottleneck and maximize the network survivability by replacing exhausted nodes at key positions with their neighbors. Neighbor switching utilizes the path redundancy of the network.

According to the three plane and five layer WSN protocol stack proposed by Akyildiz et al. [3], many existing protocols in networking layer offer solutions on how to set up or find optimal routes, but give much less effort to keep routes optimal in the viewpoint of energy. Usually maintaining established routes is common practice. On the contrary, this chapter concentrates on how to adjust established routes to balance energy dissipation. Our approach may be applied to any route setup protocols. For convenience, we only use Directed Diffusion in simulation.

We assume that in the long run all sensor nodes have approximately equal chance to become a data source. Under this assumption the network will have

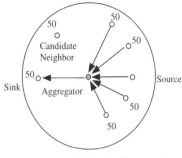 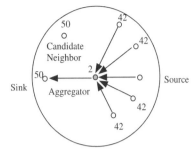

a. Initial energy 50 units for all sensor nodes   b. Residual energy after 80 time units

Fig. 6.3. Demonstration of unevenness in energy dissipation; sensing interval = 10, transmission energy = reception energy = 1.

maximum lifetime if all sensors approximately consume energy at a same rate. Like the Directed Diffusion, we also assume that nodes in the network are static. We also follow the format of task, interest and gradient in the Directed Diffusion. Now let's introduce two new concepts — neighbor switching and path rerouting.

6.2.2 Neighbor Switching

Neighbor switching, as the name implies, is to substitute one node with one of its neighbors outside the original path according to a given criterion. Neighbor switching is the first and essential step in the energy equivalence routing. Another step is path rerouting, which is discussed in the next section. Neighbor switching changes the path and overall topology at the smallest scale, so the energy uniformization may be achieved with least energy cost. In the smallest scale, neighbor switching involves only four nodes, that is, the replaced node, its preceding node, its succeeding node, and the replacing node.

The key of neighbor switching is to find the right neighbor to replace current node. It may be implemented by various algorithms, for example, replacing with the neighbor having most energy residue, or replacing central node in three successive nodes with two common neighbors. The former is used by the Shortest and Longest algorithms. The latter is called the Common Neighbor Switching, which is, according to our simulation, the most efficient in the energy equivalence routing family.

Neighbor switching can be used in other applications as well. One example is failure recovery, that is, to find a neighbor to replace the failed node. For this application, Common Neighbor Switching Algorithm could be adopted directly with 0 be both the energy of replaced node and difference threshold, see Section 6.3.5.

Neighbor switching is not necessarily limited to one hop neighbors, it could be up to k hop neighbors, where k is the diameter of the network. With more

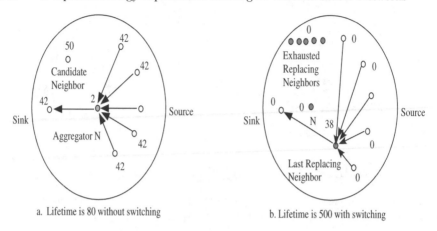

a. Lifetime is 80 without switching   b. Lifetime is 500 with switching

Fig. 6.4. Lifetime of a single task network is prolonged via neighbor switching; Sensing interval = 10, energy cost for both transmission and reception = 1.

energy dissipation in finding the optimal neighbor, replacing node could be found with better gain of the metrics. As the energy equivalence routing, the tradeoff is between the prolongation due to the uniformization and the additional energy cost.

6.2.3 Path Rerouting

Path rerouting links an outside node, which is a substitute neighbor in our algorithms, back to the original path. The energy cost of path rerouting, like in path set up, is proportional to the distance between source and destination and the scale of network topology. It is much more than neighbor switching. Similarly, much more nodes are involved in path rerouting than in neighbor switching. Neighbor switching and path rerouting are tightly coupled. The implementation in one step would affect implementation in another step.

In energy equivalence routing context, there are two path rerouting methods, namely, *shortest rerouting*, and *longest rerouting*. The shortest rerouting links the replacing neighbor to the nearest descendent node, while the longest rerouting links to the farthest descendent node. To avoid the energy waste in the chain reaction in a series of connected nodes, we define the nearest descendent node as the first descent in the path which does not need to be switched. It is better to reject the new path which intersects with original path at a non-rerouted node other than the source and destination.

In our algorithms rerouting procedure follows the Directed Diffusion protocol, with the constraint that every node on path should need no rerouting, that is, it does not have a neighbor with energy difference beyond the threshold. However, the path may contain other nodes on the original routing tree, so the new path is not necessarily disjoint with the original path. Nodes immediately following it on the original routing tree also need to set up new paths to its replacing neighbor.

6.3 EER Algorithms

In this section three algorithms for Energy Equivalence Routing (EER) are presented. For convenience, all three algorithms use Directed Diffusion to setup the routing tree. It can be regarded as a fusion of several non-branching paths from sources to sink.

Although all algorithms are distributed in nature, with communications limited to neighbors and via packet exchange, the general control framework is centralized. Every execution of EER algorithms starts at the sink, and then goes through the routing tree in a chain action, just like falling blocks in a domino game, or the spread of a wave. The frequency of the chain action is decided by the sink.

Among the three approaches Common Neighbor Switching algorithm is divergent from the other two. It is totally implemented by neighbor switching, while others by both neighbor switching and path rerouting. Common Neighbor Algorithm has the minimal time complexity and minimal energy cost, since it only processes one hop neighbors and never uses flooding style path searching, the greedy energy eater. Although there is one more nested iteration in its neighbor switching to calculate the intersection set, only a constant overhead is added given a bounded density of WSNs. This has been supported by the simulation results in Section 6.4.

The Shortest and Longest Rerouting approaches use same neighbor switching procedure and only differ in destination selection in path rerouting. For the Shortest Rerouting Algorithm and the Longest Rerouting Algorithm, the routing tree is divided into straight line segments which are bounded by converging nodes. We name them *rerouting line segments*. The major difference between the Shortest Rerouting and the Longest Rerouting is the way they determine their rerouting sections on the same line segments. The Shortest Rerouting Path Algorithm divides each segment into as many sections as possible, with each section to be shortest to minimize the topology variation. On the other hand, the Longest Rerouting Path Algorithm chooses the longest possible replacing series of successive nodes in a hope to find a better path to equalize the energy dissipation.

In all three algorithms, the source nodes and the sink node are bottle neck of the lifetime, because they can not be switched. Which node can be sources has been fixed by the interest of the task; sink is also fixed by the task. However, in many cases sink nodes are easily accessed base stations which enable energy replenishment and so have endless energy. We differentiate these two situations by a Boolean variable in our simulation.

This section gives detailed specifications of various functions, procedures, and packets used, concise explanation of three EER protocols, and formal description of algorithms.

6.3.1 Assumptions

Most of our assumptions are about the network initialization, in which the network topology is discovered and basic parameters are set. Besides the topology used in routing, which is mainly about directed paths, primary topology before the formation of paths is made of neighbors' information, if the radio radius is fixed and identical for all sensor nodes. We assume all nodes obtain important data about all of their neighbors after network initialization, like energy, position and ID.

We assume each node on the original path has a pointer called child to its preceding nodes. Parent, a set of pointers to its succeeding nodes, is optional. Particularly for Shortest Rerouting Path Algorithm and Longest Rerouting Path Algorithm, we assume that there exists a local node Ids scheme such that necessary routing information source and destination could be passed along the original path. We also assume there are network wide Ids for a WSN.

We use wake up mechanism to save the lavish energy dissipation in active listening. When hops are short in multi-hop communication, keeping in listening dissipates equal or more energy than transmitting. Wake up mechanism saves much receiving energy by turning off active listening. Synchronization is implemented using a paging channel with negligible energy dissipation. [8]

The rerouting procedure postpones the current data transfer. In EERS, when sensing interval of a task is big as in usual applications and the rerouting interval is set long enough, this delay is trivial. On the other hand, the delay may be a draw back when network is dense or large. It may not be neglected when there are many nodes in network, and the original path is long.

6.3.2 Procedures and Functions

Every algorithm needs to call following sub procedures.

- **E(A)**: returns the current energy residue of node A.
- **neighbors(A)**: returns neighbors of node A.
- **nodesInTree(T)**: returns all nodes in routing tree T.
- **directedDiffusion($nodes$, $sink$, $sources$)**: complete Directed Diffusion [5] and set up paths from source nodes to sink. In input parameters for DD, $nodes$ is a collection of nodes in the network, which represents network topology and is preestablished in network initialization; $sources$ is a collection of source nodes. We assume that neighbor set of each node is identified in initialization and only change when a neighbor runs out of energy. The output is the routing tree, a reversed directed tree. The routing tree consists of the reinforced paths in Directed Diffusion. One path in the tree is called an *original path*.

- **wake(A, B)**: node A wakes up its neighbor B or neighbors in list B.
- **sleep(A, B)**: node A puts its neighbor B or neighbors in list B into sleep state.
- **send(A, B, *packet*)**: node A sends *packet* to its neighbor B or neighbors in list B.
- **receive(A, B)**: node A receives a *packet* to its neighbor B. The function returns the packet received. If no packet is received, null is returned.
- **commonNeighborSet(N, A, B, *tree*, T)**: Node N calculates the energy-advanced common neighbor set of A and B. Let C be the common neighbor set of A and B. The function returns D, a subset of C, such that: 1) D excludes nodes on the routing tree *tree* to which A and B belong. It returns; 2) for any node $N \in D$, $E(N) \geq E(A)+T$, where T is a predefined threshold. It returns an empty set if no such element of C can be found.
- **maxENode(S)**: return a node $M \in$ node set S such that for any node $N \in S$, $E(M) \geq E(N)$.
- **multicast(A, B_1, B_2, ..., B_n, packet)**: A multicast *packet* to B_1, B_2, ..., B_n.
- **inRadius(A, B, *radius*)**: judge if nodes A and B are in the range of *radius*, in which they communicate via radio.
- **rerouting(*nodes*, *tree*, *localSink*, *localSource*)** rerouting procedure used in EERS and EERL, which generates replacing path from a single source (not a set of sources as in Directed Diffusion) to the sink. The two ends of new path will take place of *localSink* and *localSource*. It follows the Directed Diffusion, but with the constraint that every node on path should need no rerouting, that is, it does not have a neighbor with energy difference beyond the threshold. However, the path may contain other nodes on the original routing tree, so the new path is not necessarily disjoint with the original path.

The succeeding nodes of *localSource* node on the original path segment need set up new paths to the replacing node of *localSource*. At *localSink*, following adjustments are performed.
2ndLast ← *localSink.parent* (only parent node of *localSink* on original path segment);
2ndLast.child ← null;
However, the update at *localSource* is not carried out in this procedure. It will be explicitly executed in EERS and EERL. After proper updates at the two ends, the original path segment is deleted from the routing tree and all nodes except those contained by new path are put into sleep. At the end all nodes on the new path except the *localSource* are also put into sleep. This procedure may cost considerable time.

- **localSink(*replacingPath*)**: return the local sink (end node) of the replacing path generated by rerouting function.

- **localSource(*replacingPath*)**: return the local source (starting node) of the replacing path generated by rerouting function.
- **test(N, S, RC)**: node N tests if the single succeeding node S has sent N a RC packet. Return true if yes, otherwise return false.

6.3.3 Formats of Packets

- **ERR**: Extended Reroute Request packet. It has a data structure similar to the interest in Directed Diffusion. It has attributes like **type** and **timestamp**, **interval**, and **expireAt** given in a regular interest. However, instead of using attributes such as **type** and **rect**, the packet uses attributes **sink** and **sources** to indicate the sources and sink for this task. It also has the attribute **neighbors**, which contains the information about all neighbors of the sending node.
- **RRR**: Regular Reroute Request packet. It is much smaller than the ERR packet used in CNS algorithm, since it does not need to contain **neighbors**. Furthermore, the packet has a **destination** attribute, which stores the current rerouting destination.
- **CNI**: Call for Neighbors Information packet. It requests the information of receiver's neighbors.
- **NI**: Neighbors Information packet. It is the reply to CNI packet and mainly consists of the **neighbors** attribute.
- **SM(*exceptionList*)**: Switch Me packet. It finalizes all modifications of the child link for all receivers. Before this packet is sent, all updates are actually only stored in sender's buffer. Upon receiving it all receivers update their parameter especially the child link accordingly. Then all receivers except nodes in *exceptionList* switch off their power and turn into the sleep state. Sender is actually removed from the routing tree, since incoming link has broken, and no data from sources would pass it.
- **RE**: Report Energy packet. The sender uses it to request energy residue of receivers.
- **ER**: Energy Residue packet, used by sender to report its current energy level to receiver.
- **RC**: Rerouting Confirm packet. It is sent by a node N needing rerouting to its preceding node P. It is not passed more than one hop. If P needs rerouting, P gives up its rerouting intention and does nothing upon receiving the packet. If P does not received a RC from N after a timeout constant, P determines that N does not need rerouting and starts rerouting procedure from its replacing node to its RRR's destination. Source nodes do not send Reroute Confirm packets.
- **NW**: No Wait packet. It is used in EERL to tell receiver that some node else has taken rerouting and so it dose not need to do rerouting any more. This packet is passed along the segment till it reaches the destination of the segment.

- **ES**: End of Segment packet. It is used in EERL and is generated by the end node of a segment like a converging node or source node. It is passed along the segment (in the original direction of the routing tree) until it reaches the first node needing rerouting, the first converging node, or sink. In latter two situations, all nodes along the path do not need rerouting.
- **Error(*message*)**: Error packet with error message.

6.3.4 EER Common Entry Algorithm

We assume that network has been initialized and routing tree for tasks from sources to sink has been set up before EER route adjustment algorithms are called. Pseudo code for initialization and an operating system like, ever-running platform is as below.

pathTree ← directedDiffusion(*nodes*, *task.sink*, *task.sources*);
//It also found all neighbor nodes for each node.
if (*pathTree* is empty) then exit;
time ← 0;
if (time MOD (*interval*) ≠ 0) exit;

Here *nodes* is the collection of nodes in the network, representing the topology of the initialized network. *task* is a task defined in Directed Diffusion [5]. *interval* is the time interval to run CNS. We assume there is a timer to tell the current time. MOD is operation for taking remainder. Last three lines set the overall exit conditions.

We also assume that the routing tree is not empty. For empty trees, no route exists; certainly there is no need for route adjustment. Actually all algorithms are activated by reroute request packets: Extended Reroute Request (ERR) packet for CNS; Regular Reroute Request (RRR) packet for EERS and EERL. Three algorithms are called after a node has received a ERR or RRR packet. It may be called by its preceding node or itself after a little delay.

EER (*nodes, task, threshold, interval, EERType, CNSOption, timeout*) {

nextNodes ← set of next nodes in *pathTree*;
for (all *M* ∈ *nextNodes*) { // Concurrently
sink wakes up*M*;
switch (*EERType*) {
case "CNS":
sink send*M* an Extended Reroute Request packet;
CNS(*M*, *CNSOption*);
case "EERS":
sink send*M* a Regular Reroute Request packet;
EERS(*M*, *threshold*, *timeout*);
case "EERL":
sink send*M* a Regular Reroute Request packet;
EERL(*M*, *threshold*, *timeout*);
}

```
    }
    }
```

EERType is a string; it could be "CNS", "EERS" or "EERL". *CNSOption* is also a string; it could be "Plain", "Alternative" or "Intermediate". It dictates how to recover when CNP_{max} and CNS_{max} can not be connected to each other.

Below is the formal pseudo-code for EER Common Entry Algorithm.

EER (*pathTree, task, threshold, EERType, CNSOption, timeout*)

```
{
    nextNodes ← set of next nodes in pathTree;
    for (all M   ∈ nextNodes) { // Concurrently
    wake(sink, M);
    switch (EERType) {
    case "CNS":
    send(sink, M, ERR);
    CNS(M, CNSOption);
    case "EERS":
    send(sink, M, RRR);
    EERS(M, threshold, timeout);
    case "EERL":
    send(sink, M, RRR);
    EERL(M, threshold, timeout);
    }
    }
    }
```

6.3.5 Common Neighbor Switching EER Algorithm (CNS)

CNS algorithm has an ERR packet received as implicit input. Any node N on the routing tree except sink and sources has a preceding node P and a set of succeeding nodes SS. Let S $\in SS$, $\{P, N, S\}$ forms a basic unit in Common Neighbor Switching. CNP is the set of common neighbors of N and P such that: 1) it excludes nodes on the original path; 2) neighbor has more energy residue than N; 3) the energy difference between CNP and N is more than a predefined threshold. Similarly CNS is a common neighbor of N and S. CNS_{max} is a CNS with max energy residue, CNP_{max} is a CNP with max energy residue. The basic idea of Common Neighbor Switching, as shown in Fig. 6.6, is to switch N using CNP_{max} and CNS_{max}. When CNP_{max} and CNS_{max} overlap original path segment S $\rightarrow$ N $\rightarrow$ P is replaced by S $\rightarrow$ CNP_{max} (or CNS_{max}) $\rightarrow$ P, otherwise, it is replaced by S $\rightarrow$ CNS_{max} $\rightarrow$ CNP_{max} $\rightarrow$ P. Below is the simplified CNS algorithm.

```
    CNS(N, CNSOption) {
    if (N   ∈ sources) return;
    for (every S   ∈ SS) {
    N sends S a Call for Neighbors Information packet;
```

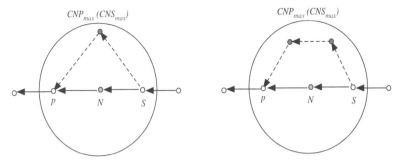

Fig. 6.5. Common Neighbor Switching.

S sends back a Neighbors Information packet;

N calculates CNP_{max}, CNS_{max};

If (CNP_{max} and CNS_{max} exist) {

if (CNP_{max}, CNS_{max} overlap) Replace $S \rightarrow N \rightarrow P$ with $S \rightarrow CNS_{max}$ (or CNP_{max}) $\rightarrow P$;

else {

if (CNP_{max} and CNS_{max} can be connected)

Replace $S \rightarrow N \rightarrow P$ with $S \rightarrow CNS_{max} \rightarrow CNP_{max} \rightarrow P$;

else Recover connection failure y according to *CNSOption*;

}

}

N sends an Extended Reroute Request packet to S and removes itself from the routing tree;

CNS(S, *CNSOption*);

}

}

Below is a more formal version of CNS algorithm.

CNS(N, *CNSOption*) {

// It has an Extended Reroute Request (ERR) packet received as implicit input, which is named ERR.

if ($N \in$ sources) return;

$P \leftarrow N$'s preceding node;

$SS \leftarrow$ set of N's succeeding nodes;

for (every $S \in SS$) {

N wakes up S and sends it a Call for Neighbors Information packet;

S sends back a Neighbors Information packet;

N calculates CNP_{max}, CNS_{max};

If (CNP_{max} and CNS_{max} exist) {

if (CNP_{max}, CNS_{max} overlap) Replace $S \rightarrow N \rightarrow P$ with $S \rightarrow CNS_{max}$ (or CNP_{max}) $\rightarrow P$;

else {

if (CNP_{max} and CNS_{max} can be connected)

Replace $S \rightarrow N \rightarrow P$ with $S \rightarrow CNS_{max} \rightarrow CNP_{max} \rightarrow P$;

else Recover connection failure y according to $CNSOption$;

   }

}

N sends an Extended Reroute Request packet to S;

N removes itself from the path tree;

Put all nodes not in use to sleep;

CNS(S, $CNSOption$);

   }

}

When N is a converging node, that is, $|SS| > 1$, old path and new path may both exist. The original data streams are shared by $S_i \rightarrow N \rightarrow P$ with $S_j \rightarrow CNS_{max} \rightarrow CNP_{max} \rightarrow P$.

Above algorithm may be adopted to virtually concurrently process each S in set SS. This will save energy and time, but require that N has adequate memory and processing capacity. In this design, N multicasts only one Call for Neighbors Information packet to all S in SS. This packet specifies the order in which Ss send back Neighbors Information packets, or this order may be set up in network initialization. Then N receives the replied Neighbors Information packets one by one. N calculates the replacing path for each branch. At last N multicast the Switch Me packets to each group of $\{P,$ CNP_{max}, CNS_{max}, $S\}$. These Switch Me packets may be integrated into a macro Switch Me packet, but the implementation costs much more energy according to the work of Bhardwaj and Chandrakasan [7], since each S in set SS has to receive all unnecessary data about other members in SS. To get the best result, our simulation follows this approach.

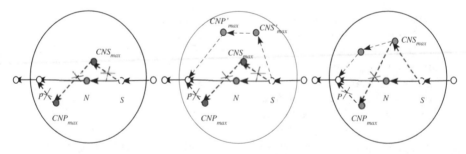

Fig. 6.6. Switching Failure Recovery.

Replacing is implemented by node's child attribute, For example, replacing $S \rightarrow N \rightarrow P$ with $S \rightarrow CNS_{max}$ (or CNP_{max}) $\rightarrow P$ is implemented as

$CNP_{max}.child \leftarrow N.child$

$S.child \leftarrow CNP_{max}$

$N.child \leftarrow$ null

Replacing $S \rightarrow N \rightarrow P$ with $S \rightarrow CNS_{max} \rightarrow CNP_{max} \rightarrow P$ is implemented as

$CNP_{max}.child \leftarrow N.child$

$CNS_{max}.child \leftarrow CNP_{max}$

$S.child \leftarrow CNS_{max}$

$N.child \leftarrow$ null

All the calculation happens on node N. There is no need of communication. At last, N multicasts a Switch Me packet to P, CNP_{max}, S, and CNS_{max}, which specifies the child link for all of them.

If CNP_{max} or CNS_{max} can not be connected (they are not neighbors), a *switching failure* occurs. Following options are provided to recover from such a failure (see Fig. 6.7):

- CNS(Plain): Use the original path $S \rightarrow N \rightarrow P$ and change nothing.
- CNS(Alternative): Find another suboptimal CNP_{max} and CNS_{max} pair which can be connected.
- CNS(Intermediate): Find an optimal common neighbor $CNPS_{max}$ of CNP_{max} and CNS_{max}, use path $S \rightarrow CNS_{max} \rightarrow CNPS_{max} \rightarrow CNP_{max} \rightarrow P$ to replace path $S \rightarrow N \rightarrow P$. N needs to wake up CNP_{max} and CNS_{max} and sends them a Call for Neighbors Information packet. CNP_{max} and CNS_{max} each send back a Neighbors Information packet to N.

Below is the formal Pseudo-code for CNS.

CNS(N, $CNSOption$) {

if ($N \in task.sources$) exit; // Exit of the recursion.

$P \leftarrow N.child$

$SS \leftarrow$ set of N's succeeding nodes

for (every $S \in SS$) {

$S \leftarrow$ a randomly selected node in SS, which has not been visited

wake(N, S);

send(N, S, CNI);

send(S, N, NI);

$CNP \leftarrow$ commonNeighborSet(N, N, P, *pathTree*, *threshold*);

// Calculation is in N, so no message needs to be sent

if (CNP is not empty) then $CNP_{max} \leftarrow$ maxENode(CNP);

else { // Simply uses the original path $S \rightarrow N \rightarrow P$.

multicast(N, P, S, SM(N, S));

send(N, S, ERR);

nodeCNS(S);

break;

}

wake(N, CNP_{max});
CNS ← commonNeighborSet(N, N, S, $pathTree$, $threshold$);
if (CNS is not empty) then CNS_{max} ← maxENode(CNS);
else {
multicast(N, P, CNP_{max},S, SM(N, S));
send(N, S, ERR);
nodeCNS(S);
break;
}
wake(N, CNS_{max});
If ($CNP_{max} = CNS_{max}$) { // They are same node.
$CNP_{max}.child$ ← $N.child$;
$S.child$ ← CNP_{max};
$N.child$ ← null;
multicast(N, P, CNP_{max} or CNS_{max}, S, SM(CNP_{max} or CNS_{max}, S));
send(N, S, ERR);
nodeCNS(S);
}
else {
if (CNP_{max} and CNS_{max} are each other's neighbor) {
$CNP_{max}.child$ ← $N.child$
$CNS_{max}.child$ ← CNP_{max}
$S.child$ ← CNS_{max}
$N.child$ ← null
multicast(N, P, CNP_{max}, CNS_{max}, S, SM(CNS_{max}, S));
send(N, S, ERR);
nodeCNS(S);
}
else {
switch ($CNSoption$) { // See Fig. 6.4
case "Plain": // Do nothing, use the original path S → N → P.
multicast(N, P, CNP_{max}, CNS_{max}, S, SM(N, S));
send(N, S, ERR);
nodeCNS(S);
break;
case "Alternative":
Find suboptimal CNP'_{max} and CNS'_{max} which can be connected;
$CNP'_{max}.child$ ← $N.child$;
$CNS'_{max}.child$ ← CNP'_{max};
$S.child$ ← CNS'_{max};
$N.child$ ← null;
multicast(N, P, CNP_{max}, CNS_{max}, CNP'_{max}, CNS'_{max}, S,
SM(CNS'_{max}, S));
send(N, S, ERR);
nodeCNS(S);

```
break;
case "Intermediate":
// Use S  → CNS_{max}  → CNPS_{max}  → CNP_{max}  →  P to replace old
// path.
send(N, CNP_{max}, CNI);
send(CNP_{max}, N, NI);
send(N, CNS_{max}, CNI);
send(CNS_{max}, N, NI);
CNPS ← commonNeighborSet(N, CNP_{max}, CNS_{max}, pathTree, threshold);
if (CNPS is not empty) {
CNPS_{max}  ← maxENode(CNPS);
wake(N, CNPS_{max});
CNP_{max}.child ← N.child;
CNPS_{max}.child ← CNP_{max};
CNS_{max}.child ← CNPS_{max};
S.child ← CNS_{max};
N.child ← null;
multicast(N, P, CNP_{max}, CNPS_{max}, CNS_{max}, S,
SM(CNS_{max}, S));
}
else multicast(N, P, CNP_{max}, CNS_{max}, S, SM(N, S));
send(N, S, ERR);
nodeCNS(S);
break;
}
}
}
sleep(N, N);
}
}
```

6.3.6 Shortest Rerouting EER Algorithm (EERS)

As mentioned in Section 6.4.1, the basic idea behind EERS is switching an over dissipated sensor node and replacing original path with shortest rerouting path. Fig. 6.7(a) illustrates EERS on a path without converging nodes. To avoid incessant and redundant rerouting, which may be very expensive in the view point of energy, a series of successive nodes in want of rerouting are group together. Only one rerouting is performed for such a series. Unlike in EERL, more than one rerouting in a single rerouting line segment is possible and encouraged in EERS. An example of EERS in a path with converging nodes is shown in Fig. 6.7(b). Actually only one more situation needs to be handled, that is, a converging node in want of rerouting. The leftmost converging node is simply the source of a rerouting line segment that is in demand of rerouting. The rightmost one is both source and sink of rerouting line segments in need of rerouting. Below is the simplified EERS algorithm.

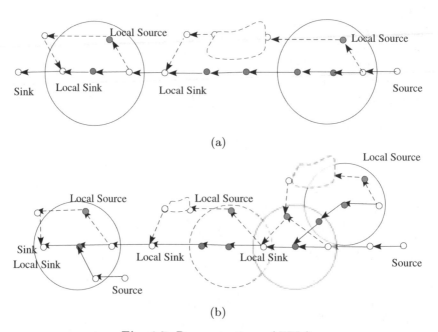

Fig. 6.7. Demonstrations of EERS.

EERS (N, *threshold*, *RCTimeout*) {
if (N ∈ sources) return;
N finds G_{max}, the neighbor with highest energy outside the routing tree;
if (E(Gmax) > E(N).+ *threshold*) { // N needs rerouting
if (N is a converging node) {
Reroute from G_{max} to RRR.destination;
RRR.destination ← N;
N switch to G_{max};
N multicasts a Regular Reroute Request packet to all S ∈ SS;
for (all S ∈ SS) EERS (S, *threshold*, *RCTimeout*); // Concurrently
}
else { // Only one S in SS
N sends a Rerouting Confirm (RC) packet to P;
N sends a Regular Reroute Request packet to S;
Wait for the Rerouting Conform packet from S until *RCTimeout* expires;
if (not received) {
Reroute from G_{max} to RRR.destination;
RRR.destination ← N;

N resends a Regular Reroute Request packet to S;
}
EERS(S, *threshold*, *RCTimeout*);
}
else { //N does not need rerouting
RRR.destination ← N;
N multicast a Regular Reroute Request packet RRR to all S ∈ SS;
for (all S ∈ SS) EERS (S, *threshold*, *RCTimeout*); // Concurrently
}
}
Below is a more formal version.

EERS (N, *threshold*, *RCTimeout*) {
// It has a Regular Reroute Request (RRR) packet received as implicit input, which is
// named RRR.
if (N ∈ sources) return;
P ← N's preceding node;
SS ← set of N's succeeding nodes;
NN ← all N's neighbors outside the routing tree;
N wakes up all G ∈ NN;
N multicasts a Report Energy packet to all G ∈ NN;
for (every G ∈ NN) G sends an Energy Residue packet to N;
G_{max} ← G ∈ NN which has highest energy;
N wakes up all S ∈ SS;
if (E(Gmax) > E(N)+ *threshold*) { // N needs rerouting
if (N is a converging node) {
Reroute from G_{max} to RRR.destination;
RRR.destination ← N;
N switch to G_{max};
N multicasts a Regular Reroute Request packet to all S ∈ SS;
for (all S ∈ SS) EERS (S, *threshold*, *RCTimeout*); // Concurrently
}
else { // Only one S in SS
N wakes up P;
N sends a Rerouting Confirm (RC) packet to P;
N sends a Regular Reroute Request packet to S;
Wait for the Rerouting Conform packet from S until *RCTimeout* expires;
if (not received) {
Reroute from G_{max} to RRR.destination;
RRR.destination ← N;
Resend a Regular Reroute Request packet to S;
}
EERS(S, *threshold*, *RCTimeout*);
}
else { N does not need rerouting

RRR.destination ← N;
N multicast a Regular Reroute Request packet RRR to all S ∈ SS;
for (all S ∈ SS) EERS (S, *threshold*, *RCTimeout*); // Concurrently
}
Put all nodes not in use to sleep;
}

This procedure must always be called after N has received the RRR packet. It may be regarded as being called by P (N's preceding Node), or by N itself after a little delay from the reception of RRR packet. In the above algorithm, the destination node for N (RRR.destination) is the nearest preceding node which has not been rerouted. If no such node exists, the destination node is the nearest preceding converging node. Upon receiving the Rerouting Confirm packet from the succeeding node, N gives up its rerouting intention and does nothing. If no Rerouting Confirm packet received till *RCTimeout*, N starts rerouting procedure from its replacing node.

Below is the formal pseudo-code for EERS.

EERS (N, $RCTimeout$) {
P ← $N.child$;
SS ← set of next nodes of N in *pathTree*;
outNodes ← neighbors(N) - nodesInTree(*pathTree*);
wake(N, *outNodes*);
send(N, *outNodes*, RE);
for (every R ∈ *outNodes*) send(R, N, RE);
replacerFound ← false;
N_{max} ← random chosen A ∈ *outNodes*;
for (every R ∈ *outNodes*) {
if ($E(R)$ ≥ $E(N)$ + *threshold*) {
replacerFound ← true;
if ($E(R)$ > $E(N_{max})$) N_{max} ← R;
}
}
if (*replacerFound*) { // N needs rerouting.
if ($|SS|$ > 1) { // N is a converging node
rerouting(*nodes*, *pathTree*, RRR.destination, N_{max});
RRR.destination ← N;
for (every S ∈ SS) {
wake(N, S);
$S.child$ ← N_{max};
multicast(N, N_{max}, S, SM(S));
send(N, S, RRR);
nodeEERS(S);
}
}
else { // N is not a converging node
S ← only element in SS;

```
wake(N, P);
send(N, P, RC);
wake(N, S);
send(N, S, RRR);
SRCTime ← 0;
repeat test(N, S, RC) until (SRCTime > RCTimeout);
if (NOT test(N, S, RC)) {
rerouting(nodes, pathTree, RRR.destination, N_max);
RRR.destination ←   N;
S.child ←   N_max;
multicast(N, N_max, S, SM(S));
}
send(N, S, RRR);
nodeEERS(S);
}
}
else { // N does not need rerouting.
RRR.destination ←   N;
for (every S   ∈ SS) {
wake(N, S);
send(N, S, RRR);
nodeEERS(S);
}
}
sleep(N, N);
}
```

6.3.7 Longest Rerouting EER Algorithm (EERL)

As mentioned in Section 6.1, EERL switches an over dissipated sensor node and replaces original path with longest rerouting path. Fig. 6.8(a) illustrates EERL on a path without converging nodes. Comparing to EERS, EERL tries to save more flooding power used in redundant rerouting. It does not skip redundant rerouting in a series of successive nodes in need of rerouting, but also combine all series of successive nodes in need of rerouting into one task of rerouting. That is a superset of all EERS rerouting paths, i.e. the farthest or longest rerouting path. Therefore at most one rerouting in a single rerouting line segment is needed in EERL. An example of EERL in a path with converging nodes is shown in Fig. 6.8(b). The leftmost converging node is simply the source of a rerouting line segment that is in demand of rerouting. The rightmost one is both source and sink of rerouting line segments in need of rerouting.

Below is the simplified EERS algorithm.

EERL (N, threshold, waitTimeout) {
if (N   ∈ sources) return;

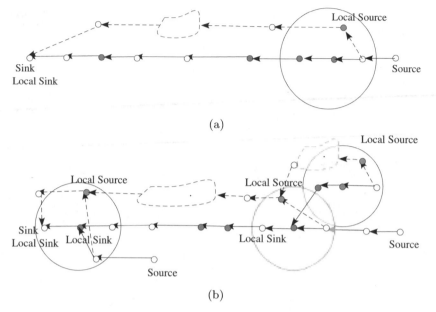

Fig. 6.8. Demonstrations of EERL.

N finds G_{max}, the neighbor with highest energy outside the routing tree;
if (N is a converging node) {
RRR.destination ← N;
N multicasts a Regular Reroute Request packet to all $S \in SS$;
for (all $S \in SS$) EERL(S, *threshold*, *waitTimeout*); // Concurrently
if (E(G_{max}) > E(N).+ *threshold*) { // N needs rerouting
N sends a No Wait packet to P, which is passed until it reaches
RRR.destination;
Reroute from G_{max} to RRR.destination;
else N sends an End of Segment packet to P, which is passed along the segment
    until it reaches the first node in want of rerouting, the first converging node,
    or sink;
    else { // N is not a converging node with only one S in SS
N sends a Regular Reroute Request packet S;
EERL(S, *threshold*, *waitTimeout*);
if (E(G_{max}) > E(N).+ *threshold*) { // N needs rerouting
Wait for an End of Segment (ES) packet or a No Wait (NW) packet from S
    until *waitTimeout* expires;
    if (ES packet received) Reroute from G_{max} to RRR.destination;
    else if (NW packet received) Give up rerouting;

else Error exit;

N sends a No Wait packet to P;

}

else {

Wait for an End of Segment packet or a No Wait packet from S until *waitTimeout* expires;

if (received) passes it to P;

}

}

}

Below is a more formal version.

EERL (N, threshold, waitTimeout) {

// It has a Regular Reroute Request (RRR) packet received as implicit input, which is

// named RRR.

if ($N \in$ sources) return;

$P \leftarrow N$'s preceding node;

$SS \leftarrow$ set of N's succeeding nodes;

$NN \leftarrow$ all N's neighbors outside the routing tree;

N wakes up all $G \in NN$;

N multicasts a Report Energy packet to all $G \in NN$;

for (every $G \in NN$) G sends an Energy Residue packet to N;

$G_{max} \leftarrow G \in NN$ which has highest energy;

Put all $G \in NN$ except G_{max} to sleep;

N wakes up all $S \in SS$;

if (N is a converging node) {

RRR.destination $\leftarrow N$;

N multicasts a Regular Reroute Request packet to all $S \in SS$;

for (all $S \in SS$) EERL(S, threshold, waitTimeout); // Concurrently

if ($E(G_{max}) > E(N)$.+ threshold) { // N needs rerouting

N wakes up P;

N sends a No Wait packet to P, which is passed along the segment until it

reaches RRR.destination;

Reroute from G_{max} to RRR.destination;

else N sends an End of Segment packet to P, which is passed along the segment

until it reaches the first node in want of rerouting, the first converging node, or sink (if a node along the path is asleep, wakes it up);

else { // N is not a converging node with only one S in SS

N wakes up S;

N sends a Regular Reroute Request packet S;

EERL(S, threshold, waitTimeout);

if ($E(G_{max}) > E(N)$.+ threshold) { // N needs rerouting

Wait for an End of Segment (ES) packet or a No Wait (NW) packet from S

until *waitTimeout* expires;

if (ES packet received) Reroute from G_{max} to RRR.destination;

else if (NW packet received) Give up rerouting;

else Error exit;

N sends a No Wait packet to P;

}

else {

Wait for an End of Segment packet or a No Wait packet from S until *waitTimeout* expires;

if (received) passes it to P;

}

}

Put all nodes not in use to sleep;

}

Like EERS, EERL is automatically called after N has received the RRR packet. It may be called by P, or by N itself after a little delay. In EERL RRR.destination is always the starting node of current rerouting line segment, and it must be a converging node or sink.

Below is the formal pseudo-code for EERL.

EERL (N, *waitTimeout*) {

$P \leftarrow N.child$;

$SS \leftarrow$ set of next nodes of N in *pathTree*;

outNodes $\leftarrow$ neighbors(N) - nodesInTree(*pathTree*);

wake(N, *outNodes*);

send(N, *outNodes*, RE);

replacerFound $\leftarrow$ false;

$N_{max} \leftarrow$ random chosen $A \in$ *outNodes*;

for (every $R \in$ *outNodes*) {

if (E(R) $\geq$ E(N) + *threshold*) {

replacerFound $\leftarrow$ true;

if (E(R) $>$ E(N_{max})) $N_{max} \leftarrow R$;

}

}

if ($|SS| > 1$) { // N is a converging node.

if (*replacerFound*) { // N needs rerouting.

wake(N, P);

send(N, P, NW);

rerouting(*nodes*, *pathTree*, RRR.destination, N_{max});

RRR.destination $\leftarrow N_{max}$;

for (every $S \in SS$) {

wake(N, S),

$S.child \leftarrow N_{max}$;

multicast(N, N_{max}, S, SM(S));

```
send(N, S, RRR);
nodeEERL(S);
}
}
else { // N does not need rerouting.
wake(N, P);
send(N, P, ES);
RRR.destination ← N;
for (every S ∈ SS) {
wake(N, S);
send(N, S, RRR);
nodeEERL(S);
}
}
}
else { // N is not a converging node.
S ← only element in SS;
wake(N, S);
send(N, S, RRR);
nodeEERL(S);
if (replacerFound) { // N needs rerouting.
waitTime ← 0;
repeat receivedPacket ←receive(N, S,) until (waitTime > waitTimeout);
if (receivedPacket is an ES packet) {
wake(N, P);
send(N, P, NW);
rerouting(nodes, pathTree, RRR.destination, N_max);
S.child ← N_max;
multicast(N, N_max, S, SM( ));
}
else if (receivedPacket is a NW packet) {
wake(N, P);
send(N, P, receivedPacket);
}
else {
send(N, S, Error("Packet Type Error or Missing Packet"));
wake(N, P);
send(N, P, Error("Packet Type Error or Missing Packet"));
exit;
}
}
else { // N does not need rerouting.
waitTime ← 0;
repeat receivedPacket ←receive(N, S,) until (waitTime > waitTimeout);
if (receivedPacket is an ES or NW packet) {
```

```
wake(N, P);
send(N, P, receivedPacket);
}
else {
wake(N, S);
send(N, S, Error("Packet Type Error or Missing Packet"));
wake(N, P);
send(N, P, Error("Packet Type Error or Missing Packet"));
exit;
}
}
}
sleep(N, N);
}
```

6.4 Simulation Analysis

6.4.1 Basic Procedure

In this section we give results of our simulation. Generally speaking, CNS shows a steady gain in lifetime, while EERS and EERL show lability and vagueness in their performance. All of them demonstrate more or less advantage over Directed Diffusion: for CNS the advantage is steady and unambiguous; for EERS and EERL, the advantage is blurry and has uncertain pattern. The performance is closely related to many factors, such as density of network, initial energy of sensors, topology, and the criteria to judge the exhaustion of network in energy. Imitating distributed computing, the simulation is actually centralized.

The performance of EER algorithms is the superposition of two opposite processes. On one hand, EER algorithms balance the energy dissipation among sensor nodes, make more paths available and prolong the network lifetime. On the other hand, frequent rerouting wastes energy while transferring to new paths, especially for flooding style EERS and EERL.

Our simulation is coded in Java 2 with Borland JBuilder 8 Enterprise IDE. It is set up as following:

1. Network Topology Generation

In a 100×100 square, we independently and at random generate n nodes with equal initial energy. These nodes are randomly distributed. Their connections are decided by their distance and a given radio radius. If two nodes are at a distance within the radius, they have an edge to connect them. As in Directed Diffusion, we assume no mobility in our simulation. [5] We assume the transmission energy, reception energy to be fixed in one run of simulation. The same topology is used for all routing protocols to be tested.

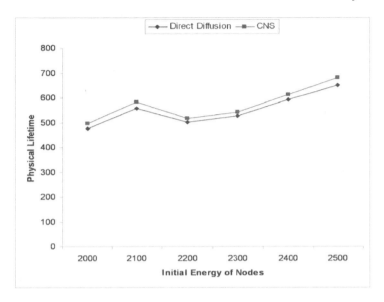

Fig. 6.9. Simulation Result of CNS (Plain): Connection Radius = 10.0, Rerouting Interval = 60, Number of Nodes = 300, Difference Threshold = 20, Rerouting Starting Period = 80, Task Generation Interval = 10, Reception Energy = 0.5, Max Time No Tasks Generated = 200, Transmission Energy = 3.0, Starting Period = 80.

2. Task Generation

Conforming to the task description in Directed Diffusion [5], tasks are generated with parameters duration and generation interval. For simplicity only one task is generated per generation interval. A 30 × 30 target area is randomly chosen by fixing its upper left corner at random. All nodes which fall in the area are source nodes. Sink is randomly chosen from all nodes.

Two strategies are employed in task generation. At first half of the development, task was generated in real time whenever it was needed. So task series for different algorithm are probably different. To balance out the possible error due to heterogeneous task series, we have to run the simulator more than 20 times and get average lifetime.

Second solution is to generate a stochastic task series with sufficient length in advance. All protocols use the same series on same topologies. This approach ensures that all tested protocols have identical input. It highly improved the credibility of results; however, it is with the price of performance of simulator. The readymade task series use a big portion of precious memory space and slow the running speed of simulator. Many times the simulator ran out of memory. The task series may be exhausted when runtime is extremely long. After frequent encountering such underflow error, we were forced to make the task list circular. All simulation results in this chapter are secured by this method.

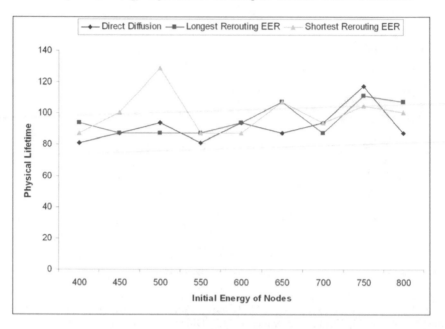

Fig. 6.10. Simulation Results of EERS and EERL: Connection Radius = 6.0, Rerouting Interval = 80, Number of Nodes = 50, Difference Threshold = 50, Rerouting Starting Period = 80, Task Generation Interval = 10, Reception Energy = 1, Starting Period = 80, Transmission Energy = 3, Task Duration = 60, Task Failure Ratio When Ending = 0.7, Sink node has same initial energy

3. Path Generation

As in Directed Diffusion, we use flooding and reinforcement to generate initial paths for a task. All protocols use exactly same path setting up procedure, since all EER algorithms use Directed Diffusion as first step. We did not assign weight for edges, so for each source node a shortest path is generated.

4. Communication and Energy Simulation

This is the major program for simulation. We use discrete time and define a time unit as the time needed to transmit a packet or receive a packet. We assume that a node transmits a received packet in the next time unit after reception. Rerouting algorithm is triggered by the rerouting interval. Rerouting paths calculation is completed discretely, imitating the real world to the life.

At any time unit, first the simulator checks if any task expires, then checks if new task generated. At last the simulator checks if it is time to call rerouting procedure. If yes, reroute along the path tree; if no, data transfer are executed. At any stop, corresponding energy cost of each node is deducted.

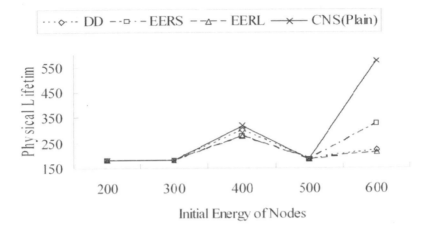

Fig. 6.11. Simulation Result with Physical Lifetime.

Connection Radius = 10.0
Rerouting Interval = 50
Number of Nodes = 150
Difference Threshold = 10
Rerouting Starting Period = 50
Task Generation Interval = 10
Reception Energy = 0.5
Max Time No Tasks Generated = 100
Transmission Energy = 2.0
Starting Period = 80
Task Failure Ratio When Ending = 0.9
Task Duration = 300
End Condition: Option 4
Sink node has endless energy
Use the same series of randomly generated tasks

6.4.2 Lifetime and End Condition

In our simulation software, two lifetimes are used. One is the *physical lifetime*, which measure the time interval of simulation from the beginning of simulation till it ends under specified end condition. Physical lifetime includes every second in simulation, even though sometime there is no active task running. And it does not matter if one hundred tasks are running or just one task is running — when they run for same interval, their physical lifetime are same. But the energy dissipation differs greatly. By this metric, in some protocol that saves and balance energy well, much more tasks may set up paths and become active than protocols with poor energy efficiency. Thus in same time, much more energy is dissipated by the good protocol, so its physical lifetime may even be shorter than the poor protocol.

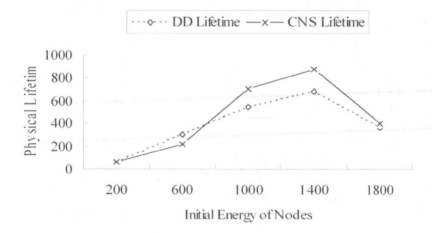

Fig. 6.12. Simulation Result with Collected Lifetime.

Connection Radius = 10.0
Rerouting Interval = 50
Number of Nodes = 150
Difference Threshold = 10
Rerouting Starting Period = 50
Task Generation Interval = 10
Reception Energy = 0.5
Max Time No Tasks Generated = 100
Transmission Energy = 2.0
Starting Period = 80
Task Failure Ratio When Ending = 0.9
Task Duration = 300
End Condition: Option 4
Sink node has endless energy
Use the same series of randomly generated tasks

Obviously the physical lifetime can not reflect energy consumption with accuracy. Therefore we use another metric called *collected lifetime*, which is the sum of the run time of all active tasks with physical lifetime as the network lifetime. However, it still has flaws. For instance, task routing trees vary remarkably in size, that is, length, nodes involved, and branches. So in same time interval, different tasks may consume wildly different energy. This is a problem that needs to be solved. We can not scale down the problem to the node level, since this way we will lose the essential global view.

The definition of network lifetime is the key for many research topics in WSNs. For a single task, it has been defined well [7], however, much effort is needed to find a definition which could offer a picture of the whole.

In our simulation, the lifetime, physical or collected, is decided by the end condition of the simulator program. We give four options. The first is the rate of failed tasks in task set, if the rate goes over a certain threshold, simulator will stop. This is the strictest option, sometimes the simulation stop when a big portion (50% ∼ 70%) of nodes have positive energy residue. Second is the rate of exhausted sensor nodes, it is looser than first, but does not always work when threshold is high. In many network topologies, especially sparse ones, some critical nodes dominate almost all possible paths; no path can be set up when they are exhausted. The third is loosest; it requires both rate of failed tasks and rate of exhausted sensor nodes be over certain threshold. Very long lifetime is reached for given threshold. The last is more practical. Under this end condition, simulator stops when the lasting time while the rate of failed tasks keeps over a given threshold is longer than another limit.

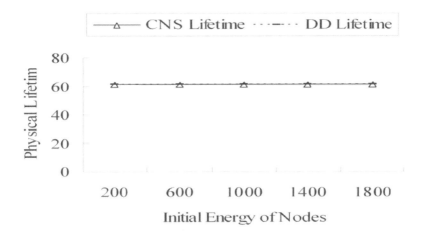

Fig. 6.13. Simulation Result with Node Number = 50

Connection Radius = 10.0
Rerouting Interval = 60
Number of Nodes = 150
Difference Threshold = 50
Rerouting Starting Period = 100
Task Generation Interval = 20
Reception Energy = 0
Max Time No Tasks Generated = 100
Transmission Energy = 3.0
Starting Period = 60
Task Failure Ratio When Ending = 0.7
Task Duration = 200
End Condition: Option 4
Sink node has endless energy
Use the same series of randomly generated tasks

6.4.3 Density of Network

In our simulation, density is controlled with two parameters. One is the number of nodes, the other is connection radius. The former is straightforward, since network area is fixed, so density is proportional to the number of nodes. The latter is not as plain as daylight. We may imagine it as a scale operation. When radius increases, each node simply enlarges its scope for neighbors, so each node will have more neighbors and more .connectivity. Fig. 6.10 shows a simple example of the first case, in which connection radius remains unchanged.

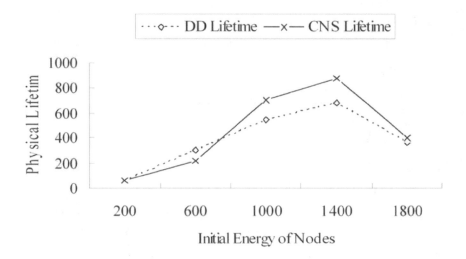

Fig. 6.14. Simulation Result with Node Number = 150
Connection Radius = 10.0
Rerouting Interval = 60
Number of Nodes = 150
Difference Threshold = 50
Rerouting Starting Period = 100
Task Generation Interval = 20
Reception Energy = 0
Max Time No Tasks Generated = 100
Transmission Energy = 3.0
Starting Period = 60
Task Failure Ratio When Ending = 0.7
Task Duration = 200
End Condition: Option 4
Sink node has endless energy
Use the same series of randomly generated tasks

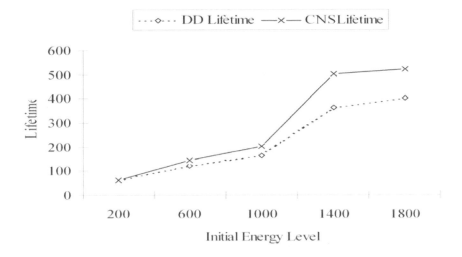

Fig. 6.15. Simulation Result with Node Number = 250
Connection Radius = 10.0
Rerouting Interval = 60
Number of Nodes = 150
Difference Threshold = 50
Rerouting Starting Period = 100
Task Generation Interval = 20
Reception Energy = 0
Max Time No Tasks Generated = 100
Transmission Energy = 3.0
Starting Period = 60
Task Failure Ratio When Ending = 0.7
Task Duration = 200
End Condition: Option 4
Sink node has endless energy
Use the same series of randomly generated tasks

From these observations, we can see that lifetime is not linear to density. Instead, most simulation results indicate that lifetime tends to decrease when density becomes too high. The rationale behind is worth further research. The possible explanation could be: (1) path setup flooding costs much more energy; (2) routing tree too big (deep and wide with much more sources; (3) too much adjusts are possible.

6.5 Conclusion

In this chapter we proposed three Energy Equivalence Routing approaches to balance the network wide energy consumption and prolong the lifetime of network. We showed that CNS is best of the three, while all of them show more or less advantage over Directed Diffusion. We defined two steps to reach the energy equivalence, that is, neighbor switching and path rerouting. A paging style wake up mechanism has been integrated. The simulation results indicate that Energy Equivalence Routing algorithms outperform the typical non-rerouting protocols. We think it is a promising approach and deserves much more future research.

Future research topics could be:

- Find effective algorithms or heuristics to determine appropriate rerouting interval and difference threshold.
- Further the research about Common Neighbor Switching Algorithm: (1) find the best parameters set up at which it gives best performance, especially the network density and initial energy; (2) investigate two remedies for connection failure —CNS alternative and CNS Intermediate, make sure if they have advantage over CNS Plain, and identify the specific circumstances in which the advantage maximizes if they do have advantage; (3) give a complexity analysis of lifetime and energy. There are many promising algorithmic details to implement ideas of Common Neighbor Switching Algorithm; we may find better ones from the family.
- Find mathematical or algorithmic proof of EER, give its upper bound and lower bound in lifetime. A concrete model should be given to quantitatively analyse the gain.
- Better the simulation software, discretise as much centralized code as possible.
- Try to find means which carry out neighbor switching and rerouting without interrupting normal data transfer. Investigate solution for possible conflict between path rerouting or neighbor switching and data transfer, and solution for conflict between rerouting of co-existing tasks.
- Explore those promising variant of current EER. For example, try the variant which only do neighbor switching and rerouting at converging nodes, because they are most liable to dissipate energy faster.

7

Time Synchronization In Wireless Sensor Networks

7.1 Introduction

Early in 1991, Jaisimha, Iyengar et al. [61] first defined some synchronization bounds for Wireless Sensor Networks based on Leslie Lamports' concepts. Römer [102] and Mills [85] in 2001-2002 pioneered the techniques and algorithms for time synchronization in sensor networks. This chapter essentially reviews these articles as the foundation for developing synchronization in Wireless Sensor Networks(WSNs). WSNs consist of large number of nodes communicating wirelessly and who have the ability of sensing. Sensor nodes need to cooperate in order to interpret the individual sensor data into a high-level result, such as integrating a time series of position measurements into a velocity estimate. Due to this the physical time at each sensor node becomes a key factor and thus time synchronization becomes one of the most important components of WSNs.

WSNs are an increasingly attractive means to bring the physical world and virtual world as close to each other as possible. They have been used in a wide variety of application areas including geophysical monitoring, transportation, military systems and business processes, WSNs are envisioned to be used to fulfill complex monitoring . But there are various challenges in the areas of the system's design. Sensor nodes are small-scale devices and as such they are very limited in the amount of energy they can store or harvest from the environment. Thus, *energy efficiency* is the major concern in a WSN. Also, for most applications many sensors need to be deployed in order to obtain the desired result and hence is also an important factor in the design of the system.

A WSN is both highly *dynamic* and *ad hoc* in nature i.e. from time to time the number of nodes may keep on changing with respect to time and changes in the surrounding environment may also interfere with the sensors operation i.e. how far the node can transmit? As such WSNs are also more susceptible to suffer failures in communication due to contention for their shared communication medium—[41] reports a message loss of 20% and above between adjacent nodes in a dense WSN. Due to the above-mentioned factors WSNS need to be self-configurable. Static configuration is unacceptable; the system must continuously adapt to make the best use of available resources.

7.2 Synchronized Time in a WSN

Time synchronization is one of the important features of all the distributed networks. In their paper, Elson and Römer present how different factors makes time synchronization particularly important as well as difficult in WSNs[35]. In this section, we will describe some of these factors: the strong connecting link between sensors and the physical world; the paucity of system energy; the need for decentralized, distributed topologies; and unpredictable connectivity. The advent of logical time [70, 82] eliminated the need for physical time synchronization in situations where only causal relationships of events are of interest to the application. But logical time is successful in only capturing the relationships between "in system" events which are defined by transmitting messages at regular intervals of time between event-generating processes. But in the case of WSN the nodes are sensing real world phenomena and hence the logical time is not sufficient, Instead physical time must be used to relate events. For example, consider the following applications:

- Object tracking: The direction, location, velocity, or acceleration of objects is determined by co-relating the sensor readings at each of the individual sensor nodes.
- State updates: The sensor node that has "sighted" the object most recently most accurately determines the current state of an object.
- Duplicate detection: Using time synchronization, nodes can determine if they are seeing two distinct real-world events, or a single event seen from two different points.
- Temporal order delivery: Also, in order to correctly interpret the result of the individual sensor node readings they need to be processed in the order of their occurrence [103]—for example, Kalman filters.

Formation of a TDMA schedule for low-energy radio operation is another example where time synchronization is needed. In this case, both transmitting and receiving signals are energy consuming operations and hence they need to be performed only when needed. One of the common techniques to conserve energy is to turn the radio off, and turning it on only when messages need to be exchanged and then turning it "off" again [96, 115]. Suppose two nodes have agreed to exchange messages (8 bits each) on the radio channel once every 60 seconds. Using a 19.2kbit/sec radio 8 bits can be transmitted in about 0.5. But, in practice, the radio must be turned "on" a little bit early to account for time synchronization error—so an expectation of even a small error will increase the total amount of time the radio is "on" i.e. the total energy that is being spent in listening to the channel. The example above demonstrates that any resource expended for synchronization reduces the resources available to perform the network's fundamental task. Also, different application can have different application requirements such as:

Energy utilization. Some synchronization schemes require extra, energy-hungry equipment (e.g., GPS receivers)

Precision. The acceptable precision might be as fine as microseconds or as coarse as seconds

Lifetime. The duration for which nodes need to be synchronized may also differ based on the application

Scope and Availability. The geographic span and coverage of nodes that are synchronized can also vary. The scope might be as small as a pair of nodes exchanging data, or as large as the entire network.

Cost and Size. These factors are one of the most important factors. If implementing the scheme requires a large amount of investment then it may not be adapted in the first place.

7.3 Traditional Network Time Synchronization

Over the years, many protocols have been designed for maintaining synchronization of physical clocks over computer networks [31, 48, 85, 116]. Mostly all these protocols rely on the same basic feature:

- A simple connectionless messaging protocol to exchange of clock information between the individual sensor nodes and the sink;
- Methods to account for the effects of nondeterminism in message delivery and processing;
- An algorithm on the sensor node to update the local clock

Mills' NTP [85] stands out by virtue of its scalability, robustness to various types of failures, self-configuration, and security in the face of deliberate sabotage, and ubiquitous deployment. But the assumptions that NTP makes are not necessarily true in sensor networks. We note some of these differences below.

7.3.1 Energy Awareness

As explained previously in Section 7.1, energy efficiency is a major concern in a WSN. These energy constraints violate a number of assumptions routinely made by classical synchronization algorithms such as

- Using the CPU in moderation is free as far as energy needs are concerned.
- Listening on the network is free.
- Occasional transmissions of messages have a negligible impact as far as energy depletion is concerned.

NTP assumes that the CPU is always available, and performs frequency discipline of the oscillator by adding small but continuous offsets to the system clock. Most of the above assumptions do not hold true in a WSN. In a low-power radio system, listening to, sending to, and receiving from the network all require spending significant amount energy and can soon lead to depletion of system energy.

7.3.2 Infrastructure Supported Vs. Ad Hoc

NTP allows construction of time synchronization hierarchies, each rooted at one of many canonical sources of external time in the Internet. The canonical sources ("Stratum 1" servers, in NTP terminology) are synchronized with each other via a variety of "out of band" mechanisms. Using this infrastructure it is possible for us to have a common view of a global timescale (UTC) to the Stratum 1 servers throughout the Internet. Consequently, nodes throughout the Internet enjoy being synchronized to a single, *global* timescale while rarely finding themselves more than a few hops away from a *local* source of this canonical time. But in case of WSNs, there may not be any external infrastructure present.

Often it is not an option to equip sensor nodes with receivers for "out of band" time references.

GPS needs high-performance digital signal processing capabilities and is thus expensive both in terms of energy consumption and component cost.

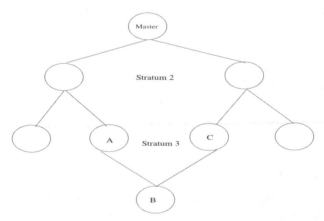

Fig. 7.1. A global timescale can lead to poorly synchronized neighbors, if the neighbors are far from the master clock and have uncorrelated loss due to divergent synchronization paths [85].

GPS also needs a line of sight to the GPS satellites—which is not available inside of buildings, etc. In this scenario, NTP-style algorithms must create a hierarchy rooted at a single node that is designated as the system's master clock. Even considering that the assumption that we have an algorithm that automatically maintains such a hierarchy in the face of node dynamics and partitions can be maintained, there is still an important problem: having only a single source of canonical time those nodes that are far away from it will be poorly synchronized to the global timescale.

This is a particularly bad situation in a WSN, where nodes closest to each other are often the ones that need the most precise synchronization—e.g.. Consider the scenario shown in Figure 7.1. Nodes A, B, and C are close to one another, but far away from the master clock. In a scheme such as NTP, B will choose either A or C as its synchronization source. Either choice will lead to poor synchronization when sharing data with the opposite neighbor. For example, if B synchronizes to C, its synchronization error to A will be quite large; the synchronization path leads all the way to the master and back. As we will discuss next, these constraints suggest that WSNs should have *no global timescale*. Instead, we propose that each node in a WSN maintain an *undisciplined* clock, augmented with relative frequency and phase information to each of its local peers.

7.3.3 Static Topology vs. Dynamics

The Internet does suffer from transient link failures; the topology remains relatively consistent from month to month, or year to year. Typically, NTP clients are manually configured with a list of "upstream" sources of time. The network dynamics in WSN do not allow such a simple kind of static configuration. In addition, the need for unattended operation of WSN prevents a manual configuration of individual nodes.

7.3.4 Connected vs. Disconnected

The dynamics in a WSN is caused due to node mobility, node failures, and environmental obstructions. This includes frequent network topology changes and network partitions. Thus information may still flow in the network but it may follow a different route every time and may have unbounded delays (depending on the movement of the node relaying the information) and are potentially unidirectional, since there might not be any nodes moving in the opposite direction. This kind of message relaying might seem like an unlikely case. However, in a sparse WSN where sensor nodes are attached to moving objects or creatures (e.g., humans, animals, vehicles, goods) or deployed in moving media (e.g., air, water) this is a major mode of communication [8, 27, 45, 72]. Grossglauser and Tse [46] even show that the communication capacity of a WSN approaches zero with increasing node density unless messages are being relayed in this way.

7.4 Design Principles for WSN Time Synchronization

Having described the shortcomings of traditional time synchronization schemes in the previous section, we can now begin to formulate requirements and new directions for time synchronization in WSNs. There are not yet any proven solutions for time synchronization in deployed WSNs. These techniques aim to build a synchronization service that conforms to the requirements of WSNs:

Energy efficiency—the energy spent synchronizing clocks should be as small as possible, bearing in mind that there is significant cost to continuous CPU use or radio listening.

Scalability—large populations of sensor nodes (hundreds or thousands) must be supported.

Robustness—the service must continuously adapt to conditions inside the network, despite dynamics that lead to network partitions.

Ad hoc deployment—time sync must work with no *a priori* configuration, and no infrastructure available (e.g., an out of- band common view of time).

7.5 Computer Clocks

Today's computing devices are equipped with a hardware oscillator assisted computer clock, which implements an approximation C(t) for real time t.

$$C(t) = \int_{t0}^{t} w(t)dt + C(t0) .$$

C(t0) is a real valued function over real time t, which depends on the angular frequency w(t) of the hardware oscillator. K is a proportional constant. For a perfect hardware clock, dC/dt would equal 1. But, all hardware clocks are imperfect and subject to clock drift. The exact drift is difficult to predict because it depends on the environmental influences (e.g. temperature, pressure, power voltage). One can usually only assume that the clock drift does not exceed a maximum value p. This means that we assume

$1 - p \leq dC/dt \leq 1 + p$.

The typical value of p achievable with today's hardware is 10^{-6}which means that the computer clock drifts away from real-time by no more than one second in ten days, which is still a significant value. Note that different computer clocks have different maximum clock drift values p_i.

7.5.1 Clock Synchronization in DSN

[61] presents the information integration algorithm.

Since the physical sensor outputs typically change as a function of time, it is necessary for each of the estimates that are integrated to be "close to each other" temporally in order for the integration process to yield meaningful results. This is achieved by time-stamping each estimate. If the estimates from different sensors are to be synchronized to obtain a final estimate at each of the commander nodes, then it is necessary that the sensors be sampled at approximately the same time.

In a distributed environment such as ours, there is no, central synchronized clock that regulates the activities of each node. Instead, each node is under the control of its own clock. Since a sensor responds to real time activities, it is convenient for the clock at each rode to provide the real, i.e., physical time. Further, since the estimates from different sensors have to be integrated, it is convenient to have the time provided by each of the sensor nodes to be close to each other. The clock at each node may not be accurate because of variety of reasons such as clock drift, variations in temperature etc. Each clock therefore has to periodically synchronize with a more accurate clock. We assume the existence of a central time server on one PE of the network that, when requested for the time at t, provides the time C(t).

Clock Behavior and Synchronization: In this section we formalize the notion of integrating abstract estimates that are temporally "close to each other" by deriving an upper bound on the time interval between the arrivals of two abstract estimates at a node. Let Cp(t) be the time provided by the clock o PE p at time t. Let Cp(t1) be the time provided by PE p be greater that C(t1), the time provided by the central time server. Let p now synchronize with the central time server. It is now possible that the new time Cp(t2) is less than Cp(t1). An abstract estimate sent out later at Cp(t2) seems as though it was sent earlier. If the abstract estimates are integrated in a non-first-in/first-out fashion at a node, this could lead to problems. Hence, we require that the time on a PE to always increase monotonically. This monotonicity could be achieved by speeding up or slowing down a PE's clock each time a correction is required on synchronizing with a central time server.

From the foregoing discussion, we can now state the following requirements for proper synchronization (the subscript p refers to PE p)

Requirement 1: Correct Time: The deviation in time of each of clock is bounded by

$$| \, t - Cp(t) \, | \quad \le \quad \varepsilon p \; (1a)$$
$$| \, t - C(t) \, | \quad \le \quad \alpha \; (1b)$$

where εp is the maximum allowable deviation in time of a clock on a PE and α is the maximum allowable deviation in time of the clock on a central time server.

Requirement 2: Correct Rate: Between resynchronizations, the drift of the clock is bounded,

$$| \, dCp(t)/dt \text{ - } 1 \, | \quad \le \quad \kappa p \; (2a) \; (2a)$$
$$| \, dC(t)/dt \text{ - } 1 \, | \quad \le \quad \sigma \; (2b)$$

where κp is the maximum allowable drift rate in time of a clock on a PE and σ is the maximum allowable drift on the clock on the central time server.

Requirement 3: The clock on each of the PE and the central time server increase monotonically.

We have assumed that the quantities εp, α, κp, σ are all fixed and are known. (These quantities could be obtained from the specifications of the manufacturer's handbook). If these quantities are time varying, then the analysis becomes very complicated. For simplicity we will assume that the constants εp is the same for all PE's and equals ε. Similarly it is assumed that $\kappa p = \kappa$. From this simplification and Requirement 1, the following inequality follows. Let Cq(t) be the time provided by the clock on PE q at time t.

Requirement 4: Synchronization bound:

$$| \, Cp(t) - Cq(t) \, | \quad \leq 2\varepsilon \quad (3)$$

Let εmin and εmax be the minimum and the maximum values of the delay in receiving the message sent by the time server to any PE. Let δmin and δmax be the corresponding values for the message sent by a PE to its neighbor. Let γ be the maximum tolerance in time t that can be integrated. This value of γ has to be derived from the sensor characteristics and the longest path between the leaf nodes and the commander node (+log n+) in our case where n is the total number of PE's.

7.6 Synchronization Algorithm

[103] presents an algorithm for time synchronization.

This algorithm uses message flows in ad hoc networks, which can be depicted by (time independent) message flow graphs, where the nodes of the graph correspond to the network nodes, each equipped with its own computer clock. Paths in the graph correspond to possibly delayed message flows between the nodes.

Using the sensor hardware present with the sensor nodes the computing nodes are able to sense events in the real world When node 1 senses an event E at real time t(E) it generates a time stamp $S_i(E)$ using its local clock, which is transmitted to other nodes using messages.

The algorithm enables all participating nodes to reason about sets of time stamps (e.g. determine the temporal ordering and time spans) received from arbitrary nodes.

Goals

- Handles all kinds of partitioning in sparse ad hoc networks
- It should not require a particular network topology beyond the one required already by the application that needs time synchronization.
- Correctness: When the algorithm claims a certain property on a set of time stamps $\{S_i(E_j) \}$ such as $\{S_1(E_1) \} < \{S_2(E_2) \}$, then this property must also hold on the corresponding set of points in real time $\{t(E_j)\}$ i.e. $\{t(E_j) \quad < \{t(E_j) \}$
- Usefulness: As far as possible it should be able to relatively order the time events i.e. there should be minimal number of maybe results
- Scalability, it should be able to support large number of sensor nodes and also changing network topology.
- Performance: Low message overhead for time synchronization.

Assumptions:

- Computer clocks with known maximum clock drift p_i.
- When an application message is exchanged between two adjacent nodes, the connection between the nodes remains established long enough to exchange another (synchronization algorithm) message between the two nodes

7.6.1 The Idea

The basic idea of the algorithm is not to synchronize the local computer clocks of the devices but instead generate time stamps using unsynchronized local clocks. Then such locally generated time stamps are passed between nodes and then they are transformed to the local time of the receiving device.

7.6.2 Time Transformation

The transformation of the real time differences Δt into computer clock differences ΔC and vice versa is the main part of the algorithm. This transformation cannot be done exactly due to the unpredictability of the computer clocks but will result in estimates (lower and upper bound). Basis of the transformation is the difference-based version of inequality 1.

$$1 - p \leq \quad \Delta C / \Delta t \leq 1 + p$$

In order to transform a time difference ΔC from the local time of one node (with p1) to the local time of a different node (with p2). ΔC is first estimated by the real time interval $[\Delta C/1+p1, \Delta C/1-p1]$, which in turn is estimated by the computer time interval

$$[\Delta C(1\text{-}p2)/(1+p1), \Delta C(1+p2)/(1+p1)]$$

(Adapted from K. Römer. Time Synchronization in Ad Hoc Networks)

7.6.3 Message Delay

Using the estimated message delay "d" that can occur when a message is sent between adjacent node the algorithm is successful in estimating the lifetime of a time stamp. It does so by following the steps outlined below:

- The sender time stamps its packet at t1 and sends it to the receiver
- On reception of this packet, the receiver time stamps it at t2 and sends it back to the sender
- The sender receives this packet and time stamps this packet at t3.
- The sender sends this packet back to receiver, which in turn time stamps it at t4.

The above steps ensure that both the sender and the receiver have an estimate about the message delay.

Using the above technique the estimated message delay d for the message M2 showed in the following figure would be

(In terms of the senders clock where ps and pr are the p values for the sender and receiver.)

$$0 \le d \le (t_5 - t4) - (t_2 - t_1)\frac{1-p_r}{1+p_s}$$

A different estimation is

$0 \le d \le (t_3 - t_2) - (t_6 - t_5)\frac{1-p_s}{1+p_r}$ (In terms of the receiver's clock that makes use of two consecutive message transfers)

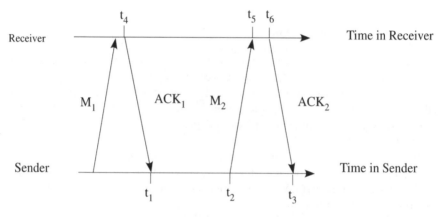

Fig. 7.2. Message delay estimation [103]

The big advantage of this second estimation is that the receiver knows estimation for d without additional message exchanges since t2 − t1 can be piggybacked on M2.

7.7 Time Stamp Calculation

We represent a time stamp $S_i(E)$ for event E in sensor node "i" by the interval $[C_{i,l}(E), C_{i,r}(E)]$ where the end points of the interval are the computer clock values relative to the computer clock in node i, where $C_i(E)$ is the value of the computer clock at real time t(E) and $C_{i,l}(E) \le C_i(E) \le C_{i,r}(E)$. Thus $S_i(E)$ is an estimation of the unknown value $C_i(E)$.

Now consider the following figure:

We have a chain of nodes 1,2,3...N and node 1 wants to pass a timestamp on to node 2,3,...N along the chain.

Each node I has 3 attributes,

- **Ri**: the local time when the message containing the time stamp is received,
- **Si**: the local time when the message containing the time stamp interval is sent and
- **Pi**: the clock drift

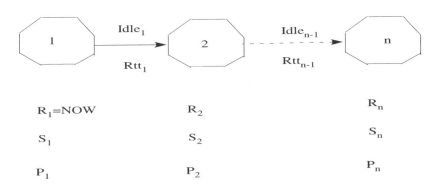

Fig. 7.3. Message Flow Graph [103]

Now let's consider the time stamp interval as it is being passed along the chain from node 1 to node N

The equations are reproduced from "K. Roemer, Time Synchronization in Ad Hoc Networks" for the sake of completeness. For more details please refer to [103].

- Node 1

$$\text{``}[r_1, r_1] = [NOW, NOW]\text{''}$$

- $$\text{``}[r_2 - (s_1 - r_1)\frac{1+p_2}{1-p_1} - rtt_1 - idle_1\frac{1-p_2}{1+p_1}),$$
$$r_2 - (s_1 - r_1)\frac{1-p_2}{1+p_1}]\text{''} \quad \text{Node 2}$$

- $$\text{``}[r_3 - (s_1 - r_1)\frac{1+p_3}{1-p_1} - (s_2 - r_2)\frac{1+p_3}{1-p_2}$$
$$-(rtt_1 - idle_1\frac{1-p_3}{1+p_1}) - (rtt_2 - idle_2\frac{1-p_3}{1+p_2}), \quad \text{Node 3}$$
$$r_3 - (s_1 - r_1)\frac{1-p_3}{1+p_1} - (s_2 - r_2)\frac{1-p_3}{1+p_2}]\text{''}$$

- $$\text{``}[r_N - (1 + p_N) \sum_{i=1}^{N-1} \frac{s_i - r_i + rtt_{i-1}}{1 - p_i} - rtt_{N-1}$$
$$+(1 - p_N) \sum_{i=1}^{N-1} \frac{idle_i}{1+p_i}, \quad \text{Node N}$$
$$r_N - (1 - p_N) \sum_{i=1}^{N-1} \frac{s_i - r_i}{1+p_i}]\text{''}$$

7.8 Improvements

The maybe results can be reduced if we maintain a history of time stamps instead of only one time stamp. Instead of updating the single time stamp upon receipt, the receiving node appends the updated time stamp together with unique node identification i and Pi.

The maybe results can also be replaced using the probability depending on the layout of the compared time stamp intervals.

7.9 Acknowledgements

The authors thank Neelay S Shah for having helped to proofread and edit this chapter.

8

Conclusions

Wireless sensor networks with battery-powered sensor nodes are intended to operate in an unattended manner for surveillance and monitoring applications. These networks should be robust to node failures as well as energy-conscious, and they should provide uninterrupted coverage and network connectivity. These objectives necessitate new network management techniques that make effective use of network resources and the available redundancy. The techniques described in this monograph have addressed the problems of effective sensing coverage, energy-efficient information processing, energy-aware self-organization, information dissemination, routing, and time synchronization. The chapters of this book have focused on a scalable and energy-efficient infrastructure for information processing in sensor networks.

First, Chapter 1 presented an overview of various networking and information processing issues related to sensor networks. Chapters 2–5 is based on work carried out at Duke University. Chapter 2 presented the virtual force algorithm (VFA) as a new approach for sensor deployment to improve the sensor field coverage after an initial random placement of sensor nodes. The cluster head executes the VFA algorithm to find new locations for sensors to enhance the overall coverage. This chapter also considered the sensor deployment problem when unavoidable uncertainty exists in precomputed sensor node locations, e.g., for airdropped sensor nodes. Inherent uncertainties in the designated sensor positions as well as the sensor field terrain have been integrated into the sensor deployment. The uncertainty-aware deployment algorithms in this chapter provide high coverage with a minimum number of sensor nodes.

Chapter 3 presented an energy-aware target localization strategy based on a two-step communication protocol between the cluster head and the sensors reporting the target detection events. This approach reduces energy consumption in the target localization process for wireless sensor networks by making use of the existing information redundancy in the target data from sensor nodes. Simulation results have shown that considerable energy is saved by using the proposed algorithm. The simulation results also illustrate the built-in advantages of reducing the communication bandwidth and filtering out false alarms.

In Chapter 5, we have described a new energy-efficient flooding algorithm termed LAF for data dissemination in wireless sensor networks. The proposed approach uses the concept of virtual grids to divide the monitored area and nodes then self-assemble into groups of gateway nodes and internal nodes. It exploits the location information available to sensor nodes to prolong the lifetime of sensor network by reducing the redundant transmissions that are inherent in flooding. This work can be extended by investigating the effect of using non-uniform grid sizes on the energy savings of LAF. Although we assumed a lossless network, LAF protocol can be easily adapted to lossy networks. A node can use the knowledge about the quality of a link to its neighbor and re-broadcast the packet multiple times. This chapter raises several interesting questions. First, we have used a uniform square grids in our simulations. However, a non-uniform grid size might be more desirable in situations where the node deployment is inherently non-uniform. Second, the size of the virtual grid can be tailored to the application and be adaptive to the activity in the network. Third, it is important to develop techniques that can dynamically reconfigure the virtual grid in a distributed manner after node failures, wearout and battery depletion.

Chapters 6–7 are based on work carried out at Louisiana State University. Chapter 6 presented three energy equivalence routing approaches to balance the network-wide energy consumption and prolong the lifetime of network. We showed that CNS is the best of the three; all three approaches improve upon the well known directed diffusion method. We defined two steps that are used to reach energy equivalence, i.e., , neighbor switching and path rerouting. A paging style wake up mechanism has also been integrated. Simulation results indicate that energy equivalence routing algorithms outperform typical non-rerouting protocols.

Finally, Chapter 7 has presented a time synchronization method for dynamic and ad hoc sensor networks. Energy efficiency and scalability are important attributes of this method.

In summary, this monograph has presented several new techniques for node deployment, energy management, self-organization, data dissemination and routing, and time synchronization in sensor networks. These techniques improve field coverage, increase network lifetime, and facilitate unattended and robust operation in harsh and hostile environments. It is expected that these techniques will lead to even more efficient protocols for scalable infrastructure design and resource management in sensor networks.

References

1. D. Agrawal, A. El Abbadi and R. Steinke, "Epidemic algorithms in replicated databases", *Proc. ACM Symposium on Principles of Database Systems*, pp. 161–172, 1997.
2. J. Agre and L. Clare, "An integrated architecture for cooperative sensing networks," *IEEE Computer*, vol. 33, pp. 106-108, 2000.
3. I. F. Akyildiz, W. Su, Y. Sankarasubramaniam and E. Cayirci, "A survey on sensor networks," *IEEE Communications Magazine*, pp. 102-114, August, 2002.
4. K. M. Alzoubi, P. J. Wan and O. Frieder, "Distributed heuristics for connected dominating sets in wireless ad hoc networks", *Journal of Communications and Networks*, vol. 4, pp. 1–8, 2002.
5. J. Albowicz, A. Chen and L. Zhang, "Recursive position estimation in sensor networks", *Proc. International Conference on Network Protocols*, pp. 35–41, 2001.
6. *"Ash Transceiver's Designers Guide"*, *http://www.rfm.com.*
7. S. Basagni, "Distributed Clustering for Ad Hoc Networks", in *Proc. ISPAN*, pp. 310–315, 1999.
8. A. Beaufour, M. Leopold, and P. Bonnet, "Smart-tag based data dissemination", *Proc. First ACM International Workshop on Wireless Sensor Networks and Applications*, 2002.
9. P. Bergamo and G. Mazzini, "Localization in sensor networks with fading and mobility", *Proc. PIMRC*, 2002 (poster).
10. University of California at Berkeley, "Mote Documentation and Development Information," *http://www.cs.berkeley.edu/~awoo/smartdust.*
11. A. Bharathidasan and V. A. S. Ponduru, "Sensor networks: an overview," *http://wwwcsif.cs.ucdavis.edu/~bharathi/.*
12. M. Bhardwaj, T. Garnett and A. P. Chandrakasan, "Upper bounds on the lifetime of sensor networks," *Proc. IEEE International Conference on Communications*, vol. 3, pp. 785-790, 2001.
13. M. Bhardwaj and A. P. Chandrakasan, "Bounding the lifetime of sensor networks via optimal role assignments," *Proc. IEEE Infocom Conference*, pp. 1587-1596, 2002.
14. K. P. Birman, M. Hayden, O. Ozkasap, Z. Xiao, S. Ni, Y. Tseng, Y. Chen, and J. Sheu, "The broadcast storm problem in a mobile ad hoc network", in

Proceedings of the Fifth Annual ACM/IEEE International Conference on Mobile Computing and Networking, pp 151-162 August 1999.

15. K. P. Birman, M. Hayden, O. Ozkasap, Z. Xiao, M. Budiu and Y. Minsky, "Bimodal Multicast", *ACM Transactions on Computer Systems*, vol. 17, pp. 41–88, May 1999.

16. D. Braginsky and D. Estrin, "Rumor routing algorithm for sensor networks", *Proc. WSNA*, 2002.

17. J. Broch, D. Maltz, D. Johnson, Y. Su and J. Jetcheva, "A performance comparison of multi-hop wireless ad hoc network routing protocols", *Proc. MOBICOM*, 1998, pp. 85-97.

18. R. R. Brooks and S. S. Iyengar, *Multi-Sensor Fusion: Fundamentals and Applications with Software*, Prentice Hall, Upper Saddle River, NJ, 1998.

19. R. R. Brooks, P. Ramanathan and A. A. Sayeed, "Distributed target classification and tracking in sensor networks", *Proceedings of the IEEE*, vol. 91, pp. 1163-1171, August 2003.

20. N. Bulusu, J. Heidemann and D. Estrin, "GPS-less low-cost outdoor localization for very small devices," *IEEE Personal Communication Magazine*, vol. 7 no. 5, pp. 28-34, 2000.

21. N. Bulusu, J. Heidemann and D. Estrin, "Adaptive beacon placement," *Proc. International Conference on Distributed Computing Systems*, pp. 489-498, 2001.

22. A. Cerpa and D. Estrin, "ASCENT: Adaptive self-configuring topologies for wireless sensor networks", *Proceedings of the IEEE INFOCOM Conference*, 2002.

23. K. Chakrabarty, S. S. Iyengar, H. Qi and E. Cho, "Grid coverage for surveillance and target Location in distributed sensor networks," *IEEE Transactions on Computers*, vol. 51, pp. 1448-1453, 2002.

24. K. Chakrabarty, S. S. Iyengar, H. Qi and E. Cho, "Coding theory framework for target location in distributed sensor networks," *Proc. International Symposium on Information Technology: Coding and Computing*, pp. 130-134, 2001.

25. G. Chartrand, *Introductory Graph Theory*, New York: Dover, 1985.

26. B. Chen, K. Jamieson, H. Balakrishnan and R. Morris, "Span: An energy-efficient co-ordination algorithm for topology maintenance in ad hoc wireless networks" , *Proc. MOBICOM*, 2001, pp. 85– 96.

27. Z. D. Chen, H. T. Kung and D. Vlah, "Ad hoc relay wireless networks over moving vehicles on highways. *Proc. MobiHoc*, 2001.

28. M. R. Clark, G. T. Anderson and R. D. Skinner, "Coupled Oscillator Control of Autonomous Mobile Robots", *Autonomous Robots*, vol. 9 no. 2, pp. 189-198, 2000.

29. T. Clouqueur, V. Phipatanasuphorn, P. Ramanathan and K. K. Saluja, "Sensor deployment strategy for target detection," *Proc. 1st ACM International Workshop on Wireless Sensor Networks and Applications*, pp. 42-48, September 2002.

30. T. Clouqueur, K. K. Saluja and P. Ramanathan, "Fault tolerance in collaborative sensor networks for target detection", *IEEE Transactions on Computers*, vol. 53, pp. 320–333, 2004.

31. F. Cristian, "Probabilistic clock synchronization", *Distributed Computing*, vol. 3, pp. 146–158, 1989.

32. B. Deb, S. Bhatnagar and B. Nath, "ReInForM: Reliable Information Forwarding using Multiple Paths in Sensor Networks", in *DCS Rutgers University Technical Report, DCS-TR-495*, March 2002.

33. S.S. Dhillon, K. Chakrabarty and S.S. Iyengar, "Sensor placement for grid coverage under imprecise detections," *Proc. International Conference on Information Fusion*, pp. 1581-1587, 2002.

34. S. S. Dhillon and K. Chakrabarty, "Sensor placement for effective coverage and surveillance in distributed sensor networks," *Proc. IEEE Wireless Communications and Networking Conference*, paper ID: TS49-2, 2003.

35. J. Elson, K. Römer, " Wireless sensor networks: A new regime for time synchronization", *ACM Computer Communication Review*, vol. 33, pp. 149–154, January 2003.

36. A. Elfes, "Occupancy grids: a stochastic spatial representation for active robot perception," *Proc. 6th Conference on Uncertainty in AI*, pp. 60-70, 1990.

37. J. Elson and D. Estrin, "Time synchraonization for wireless sensor networks," *Proc. International Parallel and Distributed Processying Symposium*, pp. 2033-2036, 2001.

38. D. Estrin, L. Girod, G. Pottie and M. Srivastava, "Instrumenting the world with wireless sensor networks," *Proc. International Conference on Acoustics, Speech, and Signal Processing* , 2001.

39. D. Estrin, R. Govindan, J. Heidemann and S. Kumar, "Next century challenges: Scalable coordination in sensor networks," *Proc. IEEE/ACM MobiCom Conference*, pp. 263-270, 1999.

40. D. Estrin, R. Govindan and J. Heidemann, *Scalable Coordination in Sensor Networks*, Technical Report 99-692, University of Southern California, 1999.

41. D. Ganesan, B. Krishnamachari, A. Woo, D. Culler, D. Estrin, and S. Wicker, "An empirical study of epidemic algorithms in large scale multihop wireless networks", Intel Research, IRB-TR-02-003, Mar. 14, 2002.

42. M.R. Garey and D.S. Johnson, *Computers and Intractability*, W.H. Freeman and Company, 1979.

43. M. Gerla and J. T. Tsai, "Multicluster, mobile, multimedia radio network", *Wireless Networks*, October 1995, vol. 1, pp. 255–65.

44. L. Girod and D. Estrin, "Robust range estimation for localization in ad hoc sensor networks", UCLA CS-TR-2000XX, 2000.

45. N. Glance, D. Snowdon and J.-L. Meunier, "Pollen: using people as a communication medium", *Computer Networks*, vol. 35, pp. 429–442, 2001.

46. M. Grossglauser and D. Tse, "Mobility increases the capacity of ad-hoc wireless networks", *Proc. INFOCOM*, 2001.

47. H. Gupta, S. R. Das, D. Johnson and Q. Gu, "Connected Sensor Cover: Self-Organization of Sensor Networks for Efficient Query Execution", *Proceedings of the Fourth Annual ACM/IEEE International Symposium on Mobile Ad Hoc Networking and Computing*, Jun 2003.

48. R. Gusell and S. Zatti, "The accuracy of clock synchronization achieved by TEMPO in Berkeley UNIX 4.3 BSD,'", *IEEE Transactions on Software Engineering*, vol. 15, pp. 847–853, 1989.

49. Z. Haas, J. Halpern and L. Li, "Gossip based ad hoc routing", *Proc. IEEE Infocom Conference*, pp. 1702–1706, 2002.

50. J. Heidemann and N. Bulusu, "Using geospatial information in sensor networks," *Proc. CSTB Workshop on Intersection of Geospatial Information and Information Technology*, 2001.

51. W. B. Heinzelman, J. Kulik and H. Balakrishnan, "Adaptive protocols for information dissemination in wireless sensor networks," *Proc. IEEE/ACM MobiCom Conference*, pp. 174-185, 1999.

52. W. B. Heinzelman, A. Chandrakasan and H. Balakrishnan, "Energy-efficient communication protocol for wireless micro sensor networks," *Proc. 33rd Annual Hawaii International Conference on System Sciences*, pp. 3005-3014, 2000.

53. W. B. Heinzelman, A. Chandrakasan and H. Balakrishnan, "An application-specific protocol architecture for wireless microsensor networks," *IEEE Transactions on Wireless Communications*, vol. 1, pp. 660-670, 2002.

54. N. Heo and P. K. Varshney, "A Distributed self spreading algorithm for mobile wireless sensor networks," *Proc. IEEE Wireless Communications and Networking Conference*, paper ID: TS48-4, 2003.

55. J. Hill, R. Szewczyk, A. Woo, S. Hollar, D. Culler and K. Piste, "System architecture directions for network sensors," *Proc. International Conference on Architectural Support for Programming Languages and Operating Systems*, 2000.

56. B. Hoffman-Wellenhof, H. Lichteneger, and J. Collins, *Global Positioning System: Theory and Practice*, Fourth Edition, Springer-Verilag, Vienna, Austria, 1997.

57. A. Howard, M. J. Matarić and G. S. Sukhatme, "Mobile sensor network deployment using potential field: a distributed scalable solution to the area coverage problem," *Proc. 6th International Conference on Distributed Autonomous Robotic Systems* , pp. 299-308, 2002.

58. C. Huitema, *Routing in Internet*, Prentice Hall, New Jersey, 1996.

59. S. S. Iyengar, L. Prasad and H. Min, *Advances in Distributed Sensor Technology*, Prentice Hall, Englewood Cliffs, NJ, 1995.

60. C. Intanagonwiwat, R. Govindan and D. Estrin, "Directed diffusion: a scalable and robust communication paradigm for sensor networks," *Proc. IEEE/ACM MobiCom Conference*, 2000.

61. D. N. Jayasimha, S.S.Iyengar and R. L. Kashyap, "Information integration and synchronization in distributed sensor networks", *IEEE Transactions on Systems, Man and Cybernetics*, vol. 21, pp. 1032–1043, 1991.

62. J. Jetcheva, Y. Hu, D. Maltz and D. Johnson, "A simple protocol for multicast and broadcast in wireless ad hoc networks", Internet Draft: draft-ietf-manet-simple-mbcast-01.txt, July, 2001.

63. M. Kokar and K. Kim, "Review of multisensor data fusion architecutres and techniques," *Proc. IEEE International Symposium on Intelligent Control*, pp. 261-266, 1993.

64. T. Kasetkasem and P. K. Varshney, "Communication structure planning for multisensor detection systems," *Proc. IEE Conference on Radar, Sonar and Navigation*, vol. 148, pp. 2-8, 2001.

65. R. Krishnan and D. Starobinski, "Message-Efficient self organization of wireless sensor networks", *Proc. WCNC*, 2003.

66. J. Kulik, W. R. Heinzelman and H. Balakrishnan, "Negotiation-based Protocols for Disseminating Information in Wireless Sensor Networks", *Wireless Networks*, 2002, Vol. 8, no 2-3, pp 169-185.

67. J. Kulik, W. Rabiner and H. Balakrishnan, "Adaptive protocols for information dissemination in wireless sensor networks", *Proc. ACM Mobicom*, pp. 174–185, 1999.

68. B. Krishnamachari and S. S. Iyengar, "Distributed Bayesian algorithms for fault-tolerant event region detection in wireless sensor networks", *IEEE Transactions on Computers*, vol. 53, pp. 241–250, March 2004.

69. T. J. Kwon, M. Gerla, V. K. Varma, M. Barton and T. R. Hsing, "Efficient flooding with passive clustering—an overhead-free selective forward mechanism for ad hoc/sensor networks", *Proceedings of the IEEE*, vol. 91, pp. 1210–1220, August 2003.

70. L. Lamport, "Time, clocks, and the ordering of events in a distributed system". *Communications of the ACM*, vol. 21, pp. 558–565, July 1978.

71. E. Lawler, J. Lenstr, A. R. Kan and D. Shmoys, *The Traveling Salesman Problem: A Guided Tour of Combinatorial Optimization*, New York: Wiley, 1985.

72. Q. Li and D. Rus, "Sending messages to mobile users in disconnected ad-hoc wireless networks. *Proc. MobiCom*, 2000.

73. Li Li and J.Y. Halpern, "Minimum-energy mobile wireless networks revisited", *Proc. IEEE International Conference on Communication*, 2001.

74. M.-J. Lin, K. Marzullo and S. Masini, "Gossip versus deterministic flooding: low message overhead and high reliability", *Proc. Int. Symp. Distributed COmputing*, pp. 253–267, 2000.

75. H. Lim and C. Kim, "Multicast tree construction and flooding in wireless ad hoc networks", *Proc. ACM Modeling, Analysis and Simulation of Wireless and Mobile Systems*, pp. 61–68, 2000.

76. S. Lindsey and C. S. Raghavendra, "PEGASIS: power-efficient gathering in sensor information systems," *Proc. IEEE Aerospace Conference*, vol. 3, pp. 1125-1130, 2002.

77. J. Liu, P. Cheung, L. Guibas and F. Zhao, "A dual-space approach to tracking and sensor management in wireless sensor networks," *Proc. 1st ACM International Workshop on Wireless Sensor Networks and Applications*, pp. 131-139, September 2002.

78. M. Locateli and U. Raber, "Packing equal circles in a square: a deterministic global optimization approach," *Discrete Applied Mathematics*, vol. 122, pp. 139-166, 2002.

79. H. Levine, W. Rappel and I. Cohen,"Self-organized in systems of self-propelled particles," *Physical Review E*, vol. 63, ID: 017101.

80. A. Mainwaring et al., "Wireless sensor networks for habitat monitoring", *First ACM Workshop on Wireless Sensor Networks and Applications*, 2002.

81. A. Manjeshwar and D. P. Agrawal, "TEEN: a routing protocol for enhanced effeciency in wireless sensor networks," *Proc. 15th International Parallel and Distributed Processing Symposium*, pp. 2009-2015, 2001.

82. F. Mattern, "Virtual time and global states in distributed systems. In *Workshop on Parallel and Distributed Algorithms*, October 1988.

83. S. Meguerdichian, F. Koushanfar, M. Potkonjak and M. B. Srivastava, "Coverage problems in wireless ad-hoc sensor networks," *Proc. IEEE Infocom Conference*, vol 3, pp. 1380-1387, 2001.

84. S. Meguerdichian, F. Koushanfar, G. Qu and M. Potkonjak, "Exposure in wireless ad-hoc sensor networks," *Proc. Mobicom Conference*, pp. 139-150, July 2001.

85. D. L. Mills, "Internet time synchronization: The network time protocol", In Z. Yang and T. A. Marsland, eds., *Global States and Time in Distributed Systems*, IEEE Computer Society Press, 1994.

86. R. Min, M. Bhardwaj, S. H. Cho, A. Sinha, E. Shih, A. Wang and A. Chandrakasan, "Low-power wireless sensor networks," *proc. VLSI Design*, 2001.

87. J. Moy, OSPF Version 2, 1991, RFC 1583, *http://www.ietf.org/rfc/rfc1583.txt*.

88. S. A. Musman, P. E. Lehner and C. Elsaesser, "Sensor planning for elusive targets," *Journal of Computer & Mathematical Modeling*, vol. 25, pp. 103-115, 1997.

89. D. C. Oppen and Y. K. Dalal, "The clearinghouse: A decentralized agent for locating named objects in a distributed environment", *ACM Transactions on Office Information Systems*, vol. 1, pp. 230–253, July 1983.

90. N. Patwari, and R. J. O'Dea, "Relative location in wireless networks", *Proc. IEEE VTC*, 1991, vol. 2, pp. 1149-1153.

91. A. Pelt, "Fault-tolerant broadcasting and gossiping in communication", *Networks*, October 1996, vol. 3, pp 143-156.

92. W. Peng and X. Lu, "On the reduction of broadcast redundancy in mobile ad hoc networks', *Proceedings of the ACM International Symposium on Mobile and Ad Hoc Networking and Computing*, pp. 129–130, 2000.

93. D. E. Penny, "The automatic management of multi-sensor systems," *Proc. International Conference on Information Fusion*, 1998.

94. Polaroid, "Technical specifications for 6500 series sonar ranging module," *http://clubweb.interbaun.com/labtop/manualweb/Polaroid/*.

95. V. Phipatanasuphorn and P. Ramanathan, "Vulnerability of sensor networks to unauthorized traversal and monitoring", *IEEE Transactions on Computers*, vol. 53, pp. 364–369, March 2004.

96. G. J. Pottie and W. J. Kaiser, "Wireless sensor networks", *Communications of the ACM*, vol. 43, pp. 51-58, May 2000.

97. N. B. Priyantha, A. Chakraborty and H. Balakrishnan, "The cricket location-support system," *Proc. IEEE/ACM MobiCom Conference*, pp. 3-43, 2000.

98. H. Qi, S. S. Iyengar, K. Chakrabarty, "Distributed sensor networks - a review of recent research," *Journal of the Franklin Institute*, vol. 338, pp. 655-668, 2001.

99. H. Qi, Y. Xu and X. Wang, "Mobile-agent-based collaborative signal and information processing in sensor networks", *Proceedings of the IEEE*, vol. 91, pp. 1172–1183, August 2003.

100. N. S. V. Rao, S. S. Iyengar, B. J. Oomen and R. L. Kashyap, "On terrain model acquisition by a point robot amidst polyhedral obstacles," *IEEE Journal of Robotics and Automation*, vol. 4, pp. 450-455, August 1988.

101. C. S. Raghavendra and S. Singh, "PAMAS: Power-aware multi-access protocol with signaling for ad hoc networks", *ACM Comm. Review*, 1998, pp. 5-26.

102. K. Roemer, "Time synchronization in ad hoc networks," *Proc. ACM Symposium on Mobile Ad Hoc Networking and Computing (MobiHoc 01)*, October 2001
www.inf.ethz.ch/vs/publ/papers/mobihoc01-time-sync.pdf.

103. K. Roemer, "Wireless sensor networks: a new regime for time synchronization", *ACM SIGCOMM Computer Communication Review*, vol. 33, pp. 149–154, 2003.

104. T. Rappaport, *Wireless Communications: Principles & Practice*, New Jersey: Prentice-Hall, Inc., 1996.

105. V. Rodoplu and T. H. Meng, "Minimum energy mobile wireless networks", *IEEE Transactions on Selected Areas in COmmunications*, vol. 7, pp. 1333–1344, August 1999.

106. J. O'Rourke, *Art Gallery Theorems and Algorithms*, Oxford University Press, New York, NY, 1987.

107. H. Sabbineni and K. Chakrabarty,"SCARE: A Scalable Self-Configuration and Adaptive Reconfiguration Scheme for Dense Sensor Networks", In S. Phoha, T.

F. La Porta and C. Griffin, ed., *Sensor Network Operations*, IEEE Press, 2005 (to be published).

108. H. Sabbineni and K. Chakrabarty, "Location-aided flooding: An energy-efficient data dissemination protocol for wireless sensor networks", *IEEE Transactions on Computers*, vol. 54, pp. 36–46, January 2005.

109. C. Schurgers, V. Tsiatsis and M. B. Srivastava, "STEM: Topology management for energy-efficient sensor networks", *Proc. IEEE Aerospace Conf.*, 2002, pp. 135–145.

110. L. Schwiebert, S.D.S. Gupta and J. Weinmann, "Research challenges in wireless networks of biomedical sensors," *Proc. IEEE/ACM MobiCom Conference*, pp. 151-165, 2001.

111. E. Shih, B. H. Calhoun, H. C. Seong and A. P. Chandrakasan, "An energy-efficient link layer for wireless micro Sensor networks," *Proc. IEEE Computer Society Workshop on VLSI*, pp. 16-21, 2001.

112. A. Sinha and A. Chandrakasan, "Dynamic power management in wireless sensor networks," *IEEE Design and Test of Computers*, vol. 18, pp. 62-74, 2001.

113. S. Slijepcevic and M. Potkonjak, "Power efficient organization of wireless sensor networks," *Proc. IEEE International Conference on Communications*, pp. 472-476, 2001.

114. K. Sohrabi and G. J. Pottie, "Performance of a novel self organization protocol for wireless ad hoc sensor networks", *Proceedings of the IEEE Vehicular Technology Conference*, 1999, pp. 1222–1226.

115. K. Sohrabi, J. Gao, V. Ailawadhi, and G. Pottie, "Protocols for self-organization of a wireless sensor network", *IEEE Personal Communications*, pp. 16–27, October 2000.

116. T. K. Srikanth and S. Toueg, "Optimal clock synchronization", *J. ACM*, vol. 34, pp. 626–645, July 1987.

117. C. Srisathapornphat, C. Jaikaeo, and C.-C. Shen, "Sensor information networking architecture," *International Workshops on Parallel Processing*, pp. 23-30, 2000.

118. R. Szewczyk et al., "Lessons from a sensor network expedition", *1st European Workshop on Wireless Sensor Networks*, 2004.

119. D. Tian and N. D. Georganas, "A Coverage-preserving node scheduling scheme for large wireless sensor networks", *Proc. WSNA*, 2002, pp. 32-41.

120. S. Tilak, N. B. Abu-Ghazaleh and W. B. Heinzelman, "A taxonomy of wireless micro-Sensor network models," *ACM Mobile Computing and Communications Review*, vol. 6, no. 2, 2002.

121. N. H. Vaidya, P. H. Krishna, M. Chatterjee and D. K. Pradhan, "A cluster-based approach for routing in dynamic networks", *ACM Computer Comm. Rev.*, Apr 1997, vol. 27, pp. 49–65.

122. P. K. Varshney, *Distributed Detection and Data Fusion*, Springer, New York, NY, 1996.

123. A. Wang, W. B. Heinzelman and A. P. Chandrakasan, "Energy-scalable protocols for battery-operated micro sensor networks," *IEEE Workshop on Signal Processing Systems*, pp. 483-490, 1999.

124. A. Wang, W. B. Heinzelman and A. P. Chandrakasan, "An energy-efficient system partitioning for distributed wireless sensor networks," *Proc. IEEE International Conference on Acoustics, Speech, and Signal Processing*, vol. 2, pp. 905-908, 2001.

125. K. Whitehouse and D. Culler, "Calibration as parameter estimation in sensor networks", *Proc. WSNA*, 2002, pp. 59-67.

126. "*Wireless Integrated Network Systems*", *http://wins.rsc.rockwell.com.*

127. Wireless LAN Medium Access Control and Physical Layer Specifications, Aug, 1999. IEEE 802.11 Standard (IEEE LAN MAN Standards Committee).

128. J. Wu and F. Dai, "Broadcasting in ad hoc networks based on self-pruning", *Proc. Infocom*, pp. 2240-2250, 2003.

129. Y. Xu, J. Heidemann and D. Estrin, "Directed diffusion: A new communication paradigm for wireless sensor networks", in *Proceedings of the ACM/IEEE International Conference on Mobile Computing and Networking*, 2000.

130. F. Ye, S. Lu and L. Zhang. GRAdient Broadcast: A Robust, Long-lived Large Sensor Network. http://irl.cs.ucla.edu/papers/grab-tech-report.ps

131. Y. Yu, R. Govindan, and D. Estrin, "Geographical and Energy Aware Routing: A recursive data dissemination protocol for wireless sensor networks", *UCLA CS Technical Report, UCLA/CSD-TR-01-0023*, May 2001.

132. Y. Xu, J. Heidemann and D. Estrin, "Adaptive energy conservating routing for multihop ad hoc routing", Technical Report 527 USC/ISI, Oct 2000.

133. Y. Xu, J. Heidemann and D. Estrin, "Geography informed energy conservation for ad hoc routing", *Proc. MOBICOM*, 2001, pp. 70–84.

134. Y. Zou and K. Chakrabarty, "Sensor deployment and target localization based on virtual forces", *Proc. IEEE Infocom Conference*, pp. 1293–1303, 2003.

135. Y. Zou and K. Chakrabarty, "Sensor deployment and target localization in distributed sensor networks", *ACM Transactions on Embedded Computing Systems*, vol. 3, pp. 61–91, February 2004.

136. Y. Zou and K. Chakrabarty, "Energy-aware target localization in wireless sensor networks," *Proc. IEEE International Conference on Pervasive Computing and Communications*, pp. 60–67, 2003.

137. Y. Zou and K. Chakrabarty, "Target localization based on energy considerations in distributed sensor networks", *Ad Hoc Networks*, vol. 1, pp. 261-272, 2003.

138. Y. Zou and K. Chakrabarty, "Uncertainty-aware sensor deployment algorithms for surveillance applications", *Proc. Globecom 2003–Next Generation Networks and Internet*, pp. 2972-2976, 2003.

139. Y. Zou and K. Chakrabarty, "Uncertainty-aware and coverage-oriented deployment for sensor networks"" *Journal of Parallel and Distributed Computing*, vol. 64, pp. 788-798, July 2004.

Index